Dedications

To my father Richard Cherry, and my mentor John Lazarchick
　　—Daniel A. Cherry

To my teachers:
　　Dr. Nedjeljko Cepic
　　Dr. Ho Huang-Chang
　　Dr. Mary M. Zutter
　　　　— Tomislav M. Jelic

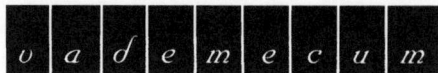

Bone Marrow
A Practical Manual

Daniel A. Cherry, MD
Laboratory Services
Trident Health System
and
Lowcountry Pathology Associates
Charleston, South Carolina, USA

Tomislav M. Jelic, MD, PhD
Charleston Area Medical Center
Charleston, West Virginia, USA

LANDES
BIOSCIENCE

AUSTIN, TEXAS
USA

VADEMECUM
Bone Marrow: A Practical Manual
LANDES BIOSCIENCE
Austin, Texas USA

Please address all inquiries to the Publisher:
Landes Bioscience, 1806 Rio Grande, Austin, Texas 78701, USA
Phone: 512/ 637 6050; FAX: 512/ 637 6079

A color version of this handbook is available in pdf at www.landesbioscience.com.

ISBN 13: 978-1-57059-708-4 (pbk)

Library of Congress Cataloging-in-Publication Data

Cherry, Daniel A. (Daniel Alan), 1961-
Bone marrow : a practical manual / Daniel A. Cherry, Tomislav M. Jelic.
   p. ; cm. -- (Vademecum)
Includes bibliographical references and index.
ISBN 978-1-57059-708-4
 1. Bone marrow--Diseases--Diagnosis--Handbooks, manuals, etc. 2. Bone marrow--Examination--Handbooks, manuals, etc. I. Jelic, Tomislav M. II. Title. III. Series: Vademecum.
 [DNLM: 1. Bone Marrow Examination--methods--Handbooks. 2. Bone Marrow Cells--pathology--Handbooks. 3. Bone Marrow Diseases--diagnosis--Handbooks. WH 39]
 RC645.7.C44 2011
 616.4'1075--dc22
                    2010053438

About the Authors...

DANIEL A. CHERRY is Medical Director of Laboratory Services for Trident Health System and is a Senior Partner of Lowcountry Pathology Associates in Charleston, South Carolina. His main interests are bone marrow pathology, surgical pathology and transfusion medicine. He completed his Doctor of Medicine degree at Spartan Health Sciences University School of Medicine, residency in Anatomic and Clinical Pathology at Marshall University School of Medicine, and fellowship in Hematopathology at the Medical University of South Carolina.

About the Authors...

TOMISLAV M. JELIC is a staff pathologist and hematopathologist in the Department of Pathology and Laboratory Medicine, Charleston Area Medical Center, Charleston, West Virginia and a Clinical Assistant Professor at the Robert C. Byrd Health Sciences Center, West Virginia University School of Medicine. His main interests include hematopathology and molecular pathology. He received his academic degrees (MD, PhD) and completed internal medicine specialization at the University of Zagreb School of Medicine, his Anatomic and Clinical Pathology residency at the Marshall University School of Medicine and Hematology fellowship at the Washington University School of Medicine, Saint Louis, Missouri.

Contents

Preface

The intentions of this manual are to familiarize beginners with the process of bone marrow evaluation, to provide a succinct preparatory review of bone marrow pathology for the pathology and clinical hematology board examinations, to remediate practitioners whose knowledge of this field is not current, and to strengthen the skills of clinicians who apply bone marrow data to the care of their patients but who do not independently examine bone marrow specimens.

The opening chapter, *Bone Marrow Basics*, is a review of information that most readers will have originally encountered in medical school. Taking nothing for granted, this material is included for the purpose of closing any possible gaps in the reader's knowledge of elementary concepts, whether from primary omission or forgetfulness. The next chapter is an overview of the principles and utilities of special studies used for the genotyping and immunophenotyping of BM. It is vital that the reader have a general understanding of these tests, since genetics and antigen expression have now equaled or surpassed the importance of morphology for bone marrow evaluation. The vast majority of this manual is devoted to hematologic neoplasms in the bone marrow and closely follows the World Health Organization classification scheme that has become the international gold standard for the characterization of these disorders. This status has been achieved by integrating all relevant data into their disease definitions, including morphology, immunophenotype, genetics and clinical features. The final two chapters of this manual address specific approaches to evaluating bone marrow that is primarily or secondarily involved by lymphoma and cytopenias due to non-neoplastic causes.

It is hoped that the impact of this manual will be some measure of demystification of the unfamiliar morphology, hi-tech ancillary studies and complex disease classification schemes that have made bone marrow pathology so intimidating to so many. Use this book as a starting-off point, and then go move some glass.

Daniel A. Cherry, MD
Laboratory Services, Trident Health System, and
Lowcountry Pathology Associates, Charleston, South Carolina, USA

Tomislav M. Jelic, MD, PhD
Charleston Area Medical Center, Charleston, West Virginia, USA

CHAPTER 1

Bone Marrow Basics

Introduction

This chapter lays the technical foundation for bone marrow (BM) examination by outlining and describing the elements of the specimen and providing a template for the BM report. Peripheral blood (PB) examination is described in some detail since the PB and BM, to some extent, form an anatomic continuum. The chapter concludes with a discussion of the possible future of BM reporting; that is, synoptic reports.

There are three major elements to the standard BM specimen and report:
- PB, including an automated complete blood count (CBC) and a manual 200 cell differential white blood cell (WBC) count
- BM aspirate, including a manual 500 cell differential count
- BM biopsy

Two additional components, touch preparations (TP) of the BM biopsy and clot sections, are sometimes included.

Bone Marrow Report Template

The following is an example of a normal BM report. This can serve as a skeleton, with minor modifications, for reporting any case.

Clinical History
Staging of follicular lymphoma.

Complete Blood Count
WBC 6.1 K/mm³ RBC 4.5 M/mm³ HGB 14.3 g/dl HCT 40.6% MCV 89.7 fL MCH 31.1 pg MCHC 34.7 g/dl RDW 12.0% PLT 245K/mm³ MPV 9.8 fl

Manual Differential Count
SEG 68% BAND 2% LYMPH 16% MONO 10% EOS 4% BASO 0%

Peripheral Blood
Microscopic examination of the peripheral blood reveals mild nonspecific anisopoikilocytosis with no significant red blood cell abnormalities. The cells are normocytic and normochromic with occasional polychromatophilic forms noted. There is no evidence of red cell agglutination or rouleaux formation. Examination of the circulating white blood cells reveals a predominance of segmented neutrophils without abnormalities of nuclear segmentation or cytoplasmic granulation. Small mature-appearing lymphocytes and morphologically unremarkable monocytes and eosinophils are also present. The estimated platelet count is normal with occasional large platelets noted.

Bone Marrow: A Practical Manual, by Daniel A. Cherry and Tomislav M. Jelic. ©2011 Landes Bioscience.

Bone Marrow Aspirate

The bone marrow aspirate is cellular and adequate for evaluation. Both the myeloid and erythroid series demonstrate cells that progress through the full range of maturation without significant morphologic abnormalities. Megakaryocytes are present in adequate numbers and demonstrate no significant morphologic abnormalities. The numbers of plasma cells and lymphocytes are not increased. An iron stain demonstrates adequate storage iron without ring sideroblasts. The following differential cell count is obtained:

SEG 10% BAND 13% LYMPH 21% MONO 2% EOS 3% ERYTHROID 20% METAMYELOCYTE 13% MYELOCYTE 10% PROMYELOCYTE 2% BLAST 1% PLASMA CELLS 2% OTHERS 0%

Bone Marrow Biopsy

The decalcified bone marrow biopsy specimen is approximately 40% cellular with an M:E ratio of 4:1. Both the myeloid and erythroid series demonstrate the full range of maturation. Megakaryocytes are present in normal numbers and demonstrate no significant morphologic abnormalities. There is no evidence of leukemia or lymphoma. There is no evidence of cells extrinsic to the marrow. The bony trabeculae are morphologically unremarkable.

Peripheral Blood Diagnosis

Morphologically normal peripheral blood smear.

Bone Marrow Diagnosis

Trilineage hematopoiesis. Negative for lymphoma.

Now that the framework has been laid, the individual pieces can be described in greater detail.

Peripheral Blood

In addition to the pathologist, the PB is examined by a laboratory technician and an automated hematology instrument. The medical technologist does a formal, microscopic 200 cell differential WBC count, observing and noting any red blood cell (RBC) morphologic abnormalities. The hematology instrument examines thousands of cells and reports numerical values, some observed directly and others calculated.

Red blood cells: The hematology instrument renders numerical values representing RBC characteristics that are listed in the upper portion of the CBC and referred to as "RBC indices":

- Mean corpuscular volume (MCV)
- Mean corpuscular hemoglobin (MCH)
- Mean corpuscular hemoglobin concentration (MCHC)
- RBC distribution width (RDW)

These values must be confirmed by microscopically observing the following RBC features:

- Cell size and variation in cell size (anisocytosis)
- Variation in cell shape (poikilocytosis)
- Cell coloration

- Specific morphologic abnormalities
- Inclusions
- Agglutination and rouleaux

Anisocytosis: Increased variation in cell size is implied by an increased RDW and can be directly observed on the PB smear. This finding may be due to a variety of cell sizes or a distinct population of cells of abnormal size such as macro-ovalocytes. If an abnormality is uniform, the RDW can be normal. A classic example is macrocytosis accompanying a myelodysplastic syndrome (MDS) with a normal RDW.

Poikilocytosis: Increased variation in cell shape is also implied by an increased RDW and may represent the presence of a variety of morphologic abnormalities. Analytic errors may occur in the RDW when there is RBC agglutination or rouleaux causing the instrument to count two or more cells clumped together as one. In general, an increased RDW occurs more often with anemias due to nutritional deficiencies (B12, folate and iron) than with anemia seen in MDS or anemia of chronic disease (ACD).

RBC size is expressed as macrocytic, normocytic or microcytic and is inferred from the MCV. On microscopic examination, RBC size should be comparable to that of the nucleus of a normal lymphocyte. The MCV may be miscalculated by the hematology instrument by measuring multiple cells stuck together as one or by measuring objects other than RBCs such as large platelets or WBC fragments. Remember that all RBC indices beginning with the word "mean" are calculated averages and can be normal despite the presence of abnormal cells. For example, in megaloblastic anemia, which is characterized by marked macrocytosis, there are frequently many small RBC fragments that can skew the MCV so that it is normal or only slightly increased. Red blood cell size is an important feature in the differential diagnosis of anemia.

RBC coloration is another important feature for the differential diagnosis of anemia and is expressed as normochromic or hypochromic. Coloration is inferred from the MCH and reflects the adequacy of hemoglobin in the RBC. On microscopic examination, the diameter of the central area of pallor should be approximately one-third of that of the RBC. Reticulocytosis can be crudely estimated by recognizing cells with a blue-gray hue to the cytoplasm polychromatophilia. The absence of an adequate reticulocyte response in proportion to an anemia is indicative of BM failure.

Abnormal RBC shapes: The hematology instrument cannot detect specific RBC morphologic abnormalities, but their presence can be hinted at by an increased RDW. These are routinely noted by the hematology technician when seen during the manual differential WBC count. A list of RBC morphologic variants is given in Table 1.1. Some laboratories quantify RBC abnormalities using the following method based on examination at 1000x magnification (10x ocular and 100x lens):

- 1(+): 3-5%/field
- 2(+): 6-15%/field
- 3(+): 16-40%/field
- 4(+): >40%/field

In general, **RBC inclusions** are related to various mechanisms of anemia, heavy metal toxicity and abnormal erythropoiesis. The spleen attempts to remove

Table 1.1. Common RBC morphologic abnormalities

Cell	Underlying Causes	Associated Diseases	Description
Acanthocyte	Accumulation of cholesterol in outer lipid bilayer of cell membrane	Abetalipoproteinemia (Bassen-Kornsweig disease), severe liver disease, anorexia nervosa, post-splenectomy	Irregular finger-like projections from cell
Bite cell (degmacyte)	Heinz body removal by spleen	Hemolytic anemia due to oxidizing drugs (Heinz body anemia), G6PD deficiency	Single semicircular defect of edge of cell
Burr cell (echinocyte, crenated cell)	Expansion of outer layer of lipid bilayer membrane of cell relative to inner layer	Artifact, uremia, liver disease, PK deficiency, bleeding ulcers, gastric carcinoma	Uniform triangular projections from edge of cell with preservation of central pallor
Ovalocyte (elliptocyte)	Altered horizontal membrane protein (spectrin-ankyrin)	Hereditary elliptocytosis, present in small numbers in normal persons	Oval or elliptical cells
Poikilocyte	Severe alteration of horizontal protein interactions	Hereditary pyropoikilocytosis, hemolytic hereditary elliptocytosis	Severe bizarre deformation of cell
Schistocyte (helmet cell, schizocyte)	Mechanical damage to cell by fibrin in microvasculature or by mechanical heart valves	Microangiopathic hemolytic anemias (DIC, TTP, HUS), prosthetic heart valves, severe burns, Zieves' syndrome, septicemia, some malignancies, cirrhosis, hemolytic anemias, renal failure, alcoholism, abetalipoproteinemia	Smaller than normal and distorted cell with one or more sharp points
Sickle cell (drepanocyte)	Aggregation of Hgb S molecules	Sickle cell anemia, Hgb SC disease (may have protruding Hgb C crystal)	Crescent-shaped cell pointed at one or both ends

continued on next page

Table 1.1. Continued

Cell	Underlying Causes	Associated Diseases	Description
Spherocyte	Deficiency of spectrin, ankyrin, and/or band 3 leading to membrane instability with loss of cell surface area; or reticuloendothelial removal of cell membrane material	Hereditary spherocytosis, immunohemolytic anemia, thermal injury, microangiopathic hemolytic anemias (DIC, TTP, HUS), toxins (clostridium, snake venom), transfusion of banked blood	Dense microcytic cell that lacks central pallor
Stomatocyte	Expansion of inner bilipid layer of cell membrane relative to outer layer, resulting in abnormal cation permeability	Hereditary stomatocytosis, vinca alkyloids, alcoholism, immunohemolytic anemia	Slit-like deformation of area of central pallor
Target cell (codocyte)	Increased ratio of cell surface area to cell volume due to excess phospholipid and cholesterol or decreased volume	Obstructive liver disease, thalassemia, Hgb C disease, Fe deficiency	Outer ring and central disk separated by area of pallor
Teardrop cell (dacryocyte)	Deformation forming a tail-like projection	Myeloproliferative disorders, myelopthisic anemia	One side of cell appears to have been pulled to a point

DIC: disseminated intravascular coagulation; HUS: hemolytic uremic syndrome; PK: pyruvate kinase; TTP: thrombotic thrombocytopenic purpura.

RBCs that contain inclusions so they are seen more often in post-splenectomy patients. Some of the more common RBC inclusions that can be observed on Wright-Giemsa staining are given in Table 1.2. Malaria and babesia may also be identified within RBCs in the PB.

Agglutination and rouleaux: RBC agglutination is the clumping of cells into irregular groups and is often due to cold antibodies. Rouleaux is the stacking of one RBC upon another, giving the appearance of stacked coins. Rouleaux is especially pronounced when caused by the presence of monoclonal protein related to plasma cell disorders but may also be seen with elevations of fibrinogen and other proteins.

Table 1.2. Inclusions seen in RBCs on Wright-Giemsa staining

Inclusion	Underlying Cause	Associated Disease	Description
Basophilic stippling	Precipitated ribosomes	Increased polychro-matophilia (fine); septicemia, thalas-semia (coarse)	Fine to coarse deep-blue granules
Cabot rings	Microtubule rem-nants from mitotic spindle	Megaloblastic ane-mia, septicemia, hemolytic anemia, post-splenectomy	Red-purple loop or figure-of-eight fila-mentous structures
Howell-Jolly bodies	Nuclear remnants	Megaloblastic anemia, hemo-lytic anemia, post-splenectomy	Smooth, round, dense basophilic structures
Pappenheimer bodies	Inorganic Fe (siderocyte) or Fe in mitochondria	Sideroblastic anemia, post-sple-nectomy, hereditary elliptocytosis	Small dense granules (fewer in number than with basophilic stippling)

White blood cells: Evaluation of the WBCs in the PB smear includes the percentages and absolute numbers of the different types and maturation stages of WBC.

Segmented neutrophils (segs, polymorphonuclear cells or PMNs) are the pre-dominant WBCs in normal PB. Variable lesser numbers of band form neutrophils (bands or stabs) may also be present and are an important reactive feature in pediat-ric patients but of little consequence in adults. Table 1.3 depicts the morphological features of normal granulocyte maturation. "Left-shift" refers to the presence of immature granulocytes in the PB. Neutrophilia with a left-shift, including up to a few blasts, may be due to almost any condition that elicits a brisk inflammatory response. Granulocytosis with cells representing the full range of maturation and a prominence of mid-level maturation (myelocytic bulge) is observed in chronic myeloid leukemia (CML). The presence of 5% or more blasts in the PB is rare with non-neoplastic conditions and should trigger a thorough investigation for clonal hematologic disease.

Neutropenia is related to important causes of morbidity and mortality, principally infection, and a common indication for BM sampling. Neutropenia is discussed in detail in Chapter 12: Examination of the Bone Marrow for Cytopenias.

Morphologic changes of neutrophils: Neutrophils may demonstrate a variety of morphologic abnormalities. The most common are reactive in nature:

- **Toxic granules:** neutrophil granules that have retained their basophilia, reflecting a lack of maturation
- **Döhle bodies:** persistent basophilia of the cytoplasm that becomes discrete
- **Cytoplasmic vacuolization:** a degenerative change that strongly correlates with sepsis

Table 1.3. Normal granulocyte maturation

Maturation Stage	Description
Myeloblast	15-20 μm; high N:C; bluish cytoplasm without granules or with azurophilic granules; cytoplasmic tags; round reddish nucleus with fine, evenly dispersed interlaced chromatin, and usually two or more nucleoli
Promyelocyte	Bluish cytoplasm now contains dark blue to reddish, purple-red primary granules and has smooth edges; N:C less than that of blasts; nucleus is round to oval and slightly coarser than blast with no indentation, inconspicuous nucleoli, and perinuclear hoff
Myelocyte	Light area of ill-defined reddish primary granules develops adjacent to nucleus; smaller than promyelocyte, with relatively more cytoplasm; oval nucleus flattened on one side with inconspicuous nucleoli
Metamyelocyte	Nucleus becomes slightly indented; cytoplasm contains small secondary pinkish-blue granules; nucleolus is absent.
Band neutrophil	Nuclear indentation becomes >1/2 the width of the nucleus
Segmented neutrophil	Nucleus is divided into separate segments connected by a thin filament; cytoplasm is light pink with numerous small light pink to blue-purple granules

N:C: nuclear to cytoplasmic ratio

When present, these findings provide a small measure of comfort that the neutrophilia is due to a reactive etiology rather than a myeloproliferative neoplasm (MPN). When accompanied by neutropenia, toxic changes may indicate Gram-negative bacterial sepsis. Similar morphologic changes may occur with rare genetic diseases such as the following:

- **Alder-Reilly anomaly** with coarse granulations resembling toxic granules
- **May-Hegglin anomaly** with Döhle bodies
- **Chediak-Higashi syndrome** demonstrating abnormal ribosomes that appear as large dark granules

Neutrophil hypersegmentation and gigantism are two of many manifestations of megaloblastic anemia. Hypersegmentation is defined as five lobes in ≥5% of the neutrophils or six or more lobes in any neutrophil. Three hereditary diseases demonstrate hypersegmentation: **hereditary hypersegmentation of neutrophils** and **hereditary giant neutrophils** are isolated, fairly benign blood disorders, while **myelokathexis** is associated with growth retardation and severe skeletal abnormalities.

Neutrophil dysplasia (dysgranulocytopoiesis) is one of the more important abnormalities that can be identified in the PB since this may be the sole morphologic abnormality in cases of MDS. Neutrophil dysplasia is described in detail in Chapter 2.

Auer rods are a form of abnormal granulation in which the granules become aligned and form rod-like structures. The presence of Auer rods is always neoplastic.

Specific granule deficiency is a hereditary disorder in which the neutrophils demonstrate hyposegmentation and hypogranulation due to the absence of secondary granules. This disease is associated with recurrent infections occurring from birth.

Leukoerythroblastosis is a granulocytic left-shift accompanied by nucleated RBCs (NRBCs). A leukoerythroblastic blood smear may be seen with BM occupying processes such as lymphoma or metastatic carcinoma and is characteristic of primary myelofibrosis.

Lymphocytes in the PB should be predominantly small and mature in appearance. Small numbers of **atypical lymphocytes** (ATLs) are commonly encountered and are usually regarded as reactive in nature, although a specific etiology is seldom known. The term "ATL" implies the absence of overt neoplastic features but leaves open the possibility of a clonal process. The appearance of ATLs varies. In some, there is increased and less dense cytoplasm and a slightly enlarged nucleus with decreased chromatin density and obvious nucleoli. Adjacent RBCs may cause indentations of the cytoplasm accompanied by increased cytoplasmic density in a scalloped pattern. Azurophilic cytoplasmic granules can be present, and sometimes there is lobation of the nucleus resembling a monocyte. This is the origin of the misnomer "infectious mononucleosis." The presence of cytoplasmic vacuoles is a useful differential feature that is frequently seen in monocytes and rarely in ATLs. In other instances, ATLs demonstrate increased nuclear and cytoplasmic density imparting a plasma cell-like appearance. Still other ATLs appear immature with cytomegaly, nucleomegaly, reticular chromatin, conspicuous nucleoli and abundant basophilic cytoplasm. These can be difficult to distinguish from the lymphoblasts of acute lymphoblastic leukemia (ALL) and are referred to as "reticular lymphocytes" or "non-leukemic lymphoblasts." Differentiation from ALL is accomplished by clinical findings, viral serology and immunophenotyping with flow cytometry (FCM).

Lymphocytosis may be relative or absolute. Relative lymphocytosis is more common and is defined as normal absolute numbers of lymphocytes but relatively increased in percentage due to neutropenia. Relative lymphocytosis is most commonly associated with viral illnesses but may also be seen with neutropenia from other causes and in Addison's disease. Absolute lymphocytosis is defined as an absolute lymphocyte count (calculated by multiplying the percentage of lymphocytes by the WBC and then dividing by 100) >4.0 x 10^9/L in adults, or >8.8 x 10^9/L in children. It is convenient to separate the non-neoplastic causes of lymphocytosis based on presence or absence of significant numbers of ATLs (see Table 1.4). **Pertussis infection** and **infectious lymphocytosis** (a contagious viral illness seen mainly in children) demonstrate virtually no lymphocyte atypia. Pertussis infection in children gives a PB smear picture similar to that seen with chronic lymphocytic leukemia (CLL) in adults.

Neoplastic lymphocytes: Lymphocytes are designated as neoplastic when they demonstrate clonality. B-cell clonality is identified by restricted expression of immunoglobulin (Ig) light chains by FCM or immunohistochemistry (IHC), or by Ig gene rearrangement studies. T-cell clonality can be identified by T-cell receptor gene rearrangement studies or inferred by an abnormal immunophenotype

Table 1.4. Common causes of absolute lymphocytosis with and without significant numbers of atypical forms

No or Minimal Numbers of Atypical Lymphocytes
- Pertussis
- Infectious lymphocytosis
- Some children with adenovirus

Significant Numbers of Atypical Lymphocytes Present
- Infectious mononucleosis
- Adult cytomegalovirus infection
- Sometimes viral hepatitis
- Toxoplasmosis
- Very early phase following severe trauma

with FCM. Clonal lymphocytes vary considerably in appearance. Lymphocytes that are morphologically indistinguishable from normal cells are seen in CLL and in some cases of childhood ALL, whereas the neoplastic cells of adult T-cell leukemia/lymphoma (ATLL) may demonstrate such bizarre nuclear features that they can be recognized as malignant on scanning power. The lymphoblasts in many cases of ALL are not morphologically discernable from the myeloblasts of acute myeloid leukemia (AML). Any lymphocytic neoplasm present in the BM can, theoretically, leak cells into the PB, including large cells such as those of anaplastic large cell lymphoma (ALCL). It should be kept in mind that not all clonal lymphocytes are malignant. Populations of clonal lymphocytes have been identified in reactive conditions. Neoplastic lymphocytes are discussed more fully in the chapters on *Evaluation of the Bone Marrow for Lymphoma* and *Mature B-, T- and NK-Cell Leukemias*.

Lymphopenia is defined as an absolute lymphocyte count <1.5 x 10^9/L in adults and <3.0 x 10^9/L in children. Lymphopenia is seen in acquired immunodeficiency syndrome (AIDS), mainly due to decreased CD4 helper T cells, but is otherwise rare occurring in genetic immunodeficiency disorders. Following myeloablative therapy, the number of lymphocytes is decreased; however, they usually account for the majority of cells present in both the PB and BM.

Monocytosis is defined as an absolute monocyte count >1.0 x 10^9/L in adults and >1.2 x 10^9/L in neonates. Monocytosis is most commonly reactive in nature and due to non-infectious causes. A number of hematologic neoplasms and rarely carcinomas may also have associated reactive monocytosis in which the monocytes are not clonal. Several chronic diseases and the recovery phase of some acute diseases are also accompanied by monocytosis. Chronic infections are a less common cause of reactive monocytosis. A somewhat lengthy list of specific disorders in which reactive monocytosis may be seen is presented in Table 1.5.

Neoplastic monocytes are seen in the PB in acute myelomonocytic leukemia (AMML), acute monocytic leukemia, chronic myelomonocytic leukemia (CMML) and histiocytic sarcoma.

Monocytopenia is defined as <0.2 x 10^9/L and is rare, occurring in the first few hours after the initiation of prednisone therapy and with hairy cell leukemia (HCL).

Table 1.5. Disorders associated with reactive monocytosis

Chronic Diseases
- Cirrhosis
- Collagen vascular diseases
- Non-tropical sprue
- Sarcoidosis
- Ulcerative colitis

Neoplastic Diseases
- Carcinomas (rarely)
- Hodgkin lymphoma
- Non-Hodgkin lymphoma
- Plasma cell myeloma

Chronic Infections
- Fungal
- Protozoal
- Rickettsial
- Tuberculosis
- Viral
- Subacute endocarditis

Miscellaneous
- Fever of unknown origin (FUO)
- Hemolytic anemias
- Idiopathic thrombocytopenic purpura (ITP)
- Post-splenectomy
- Recovery from acute infection or agranulocytosis
- Unexplained

Eosinophils are normally seen in small numbers in the PB and show a slight diurnal variation that parallels that of cortisol levels, being lower in the morning and higher in the afternoon. Because they are present in such small numbers, some laboratories establish separate normal ranges depending on whether the count is automated or manual. Normal eosinophils are spherical with red cytoplasmic granules. The nucleus is usually bi-lobed, although occasional tri-lobed forms may be seen. Frequent tri-lobed forms or those with more lobes are abnormal and may become prominent in CML, as are cells with hypogranulation or cytoplasmic vacuolization.

Eosinophilia is defined as an absolute eosinophil count of >0.6 x 10^9/L and may be due to a number of causes, most commonly allergic and hypersensitivity reactions. Table 1.6 lists some of the conditions in which eosinophilia may occur. **Eosinopenia** results from adrenal glucocorticoid and epinephrine release and, therefore, may occur in any condition of severe stress like trauma or burns. Eosinopenia is rarely identified because such small numbers of eosinophils are normally present in the PB.

Basophils are spherical cells with prominent dark blue to purple granules that partially or completely obscure the nucleus. Basophils are present in the PB in very small numbers and, similar to eosinophils, some laboratories establish separate

Table 1.6. Common conditions associated with eosinophilia

Allergic Diseases
- Bronchial asthma
- Other allergic and hypersensitivity reactions
- Seasonal rhinitis

Skin Disorders
- Acute urticarial reactions
- Atopic dermatitis
- Eczema
- Pemphigus

Parasitic Infestations
- Cysticercosis
- Tapeworm
- Trichinosis
- Visceral larva migrans

Bacterial Infections
- Scarlet fever
- Chorea

Pulmonary Eosinophilias
- Löffler's syndrome
- Pulmonary infiltrates with eosinophilia (PIE)
- Tropical eosinophilia

Hypereosinophilic Syndrome
Hematologic Neoplasms
- Acute lymphoblastic leukemia (ALLeo)
- AMML
- Chronic eosinophilic leukemia
- Hodgkin lymphoma
- Myeloproliferative neoplasms (especially CML)

Miscellaneous
- Other neoplasms
- Ovarian cysts
- Pernicious anemia
- Post-splenectomy
- Recombinant human interleukin therapy
- Sarcoidosis

reference values depending on whether a manual or automated differential cell count is done.

Basophilia is defined as an absolute basophil count of >0.2 x 10^9/L and can be seen as a reactive finding with a number of disorders (Table 1.7), especially allergic and hypersensitivity conditions. Basophilia may also occur with AML and MPNs, particularly CML. Acute myeloid leukemias with basophilia are associated with the del(12)(p11-p13) chromosomal abnormality. Acute basophilic leukemia is an extraordinarily rare disease. **Basopenia** may be seen with long-term glucocorticoid

Table 1.7. Disorders associated with basophilia

- Allergic disorders and hypersensitivity reactions
- Carcinomas (rare)
- Chicken pox
- Chronic hemolytic anemia
- Chronic renal disease
- Collagen vascular diseases
- Endocrinopathy
- Influenza
- Irradiation
- Medications (estrogen, anti-thyroid agents)
- Myeloproliferative neoplasms (especially CML)
- Post-splenectomy
- Small pox
- Ulcerative colitis

therapy, acute infection or stress, and in some patients with hyperthyroidism, but is difficult to define and identify, since such small numbers are normally present.

Platelets

Platelet counting: Platelets are difficult for hematology instruments to count because they are very small and sticky, tending to adhere to glass, debris and each other. Therefore, a low platelet count derived from the hematology instrument must be confirmed by direct observation of the PB smear. Clumping and satellitism (adherence of platelets to RBC or neutrophil membranes) are two common causes of error in automated platelet counts and are self-evident on examination of the smear. Collection of the specimen in a heparin tube corrects these artifacts. To estimate the platelet count on the PB smear, choose an area where the platelets are most numerous but have not yet begun sticking to each other, count them in one-quarter of a 500x field (50x lens and 10x ocular) and multiply by 10. For example, if 12 platelets are counted in this fashion, then the estimated platelet count is 120×10^9/L (an alternate method is described in the chapter on cytopenias).

Platelet morphology can occasionally provide a clue to platelet disorders. The most common abnormality is large or giant platelets (Table 1.8). Hypogranular platelets occur most often with MPNs or, rarely, due to hereditary disorders of platelet granules.

Bone Marrow Aspirate

Adequacy: A general statement can be made about the BM cellularity based on the appearance of the aspirate, but a more accurate estimate is done on the biopsy. Adequacy of the BM aspirate is not determined by the percentage cellularity, as in the biopsy, but by the presence of BM particles, commonly referred to as spicules. The BM can be hypercellular but inadequately aspirated due to fibrosis or tightly packed malignant cells, thus, yielding few spicules and inaccurately reflecting the BM content. If the BM is hypocellular, the aspirate will appear paucicellular but with spicules, indicating that the cellularity of the smear represents that of the marrow.

Table 1.8. Disorders associated with large or giant platelets

- Bernard-Soulier syndrome
- Idiopathic thrombocytopenic purpura (ITP)
- Myelofibrosis
- MPNs (also may be hypogranular)
- Myelophthisis
- Post-splenectomy
- Thrombasthenia

Types of cells seen in the aspirate: For the purposes of evaluation and description, it is useful to divide the cells found in the BM aspirate into four broad groups:
- Myeloid
- Erythroid
- Megakaryocytic
- Lymphoid and plasma cells (PCs)

Granulocytic, erythroid and megakaryocytic cells are all myeloid in origin, but in the BM report "myeloid" is commonly used to refer to the granulocytic cells. Extrinsic cells (those not normally seen in the BM such as metastatic carcinoma cells) are best seen in the biopsy specimen and are usually not identified in the aspirate. The BM aspirate is the optimum specimen for evaluation of hematopoietic cell morphology and maturation.

Myeloid maturation: Referring back to Table 1.2, it is apparent that there are easily recognized morphologic features that identify the stages of maturation of individual granulocytic cells. The 500 cell differential count may be compared to broad normal ranges for maturational abnormalities. Reactive changes and the administration of growth factors such as colony stimulating factor-granulocyte (CSF-G) are non-neoplastic etiologies for increased immature granulocytes (left-shift). More important, however, are neoplastic causes such as MDS and AML. Myeloid maturation arrest with few or no mature granulocytes occurs in some cases of AML.

Morphologic abnormalities of the myeloid cells include hypersegmentation, hyposegmentation, gigantism and the presence of Auer rods. These features are discussed more fully in the chapters on MDS and AML. **Increased numbers of eosinophils** and eosinophilic precursor cells may be either reactive or neoplastic and may be accompanied by abnormalities of nuclear and/or granulation. Rarely, **increased basophils or mast cells** occur as either a reactive phenomenon or a neoplasm, as in acute basophilic leukemia or systemic mast cell disease.

Erythroid maturation: The delineation of the specific maturation stages of the erythroid cells is less practical than with granulocytes because the transition from one stage to the next is more subtle and lacking convenient morphologic milestones. Instead, all erythroid precursor cells are lumped together in the differential count. Whether or not their maturation is orderly is determined by identifying cells that are in the early, mid- and late stages of maturation. Early erythroid cells are similar in appearance to early myeloid cells but have more round nuclei with denser linear chromatin. The presence of hemoglobin in the cytoplasm imparts a reddish, royal-blue color that is quite distinct. As they mature, the nuclei become smaller and darker with increased chromatin density, and the cytoplasm contains a higher proportion of red staining due to increased hemoglobin content. The late normoblast, the last stage of maturation before the nucleus is extruded, demonstrates a small, round, dense nucleus with red cytoplasm that is similar in color to that of mature RBCs in the PB with only the slightest light-blue tint. In practice, subtle abnormalities of erythroid maturation are not usually evident.

Morphologic abnormalities of the erythroid precursor cells usually take the form of nuclear membrane and shape irregularities, multinucleation and/or megaloblastoid changes. These findings are described fully in the chapter on MDS.

Megakaryocytes: A general estimation of the number of megakaryocytes may be made from the aspirate and stated as adequate or inadequate. A more accurate estimate is done on the biopsy specimen as will be subsequently described. Morphologic abnormalities of megakaryocytes are best seen in the aspirate and include macromegakaryocytosis, micromegakarycotosis, nuclear hyperlobation and hypolobation, and complete separation of nuclear segments. Again, these abnormalities will be fully described in the chapter on MDS.

Plasma cells are easily identified by their characteristic appearance, consisting of an oval or round eccentrically placed nucleus, abundant blue cytoplasm, paranuclear hoff (pale area in the cytoplasm) and frequent nuclear and cytoplasmic inclusions. Demonstration of Ig light chain monotypism by IHC (biopsy) or FCM (aspirate) indicates clonality and is crucial when evaluating plasmacytosis, since there is much overlap in the number of PCs seen with reactive and neoplastic conditions. Furthermore, even advanced cases of plasma cell myeloma usually demonstrate little atypia.

Lymphocytes: Increased lymphocytes can be identified in the BM aspirate, but they tend to deceptively blend with the other BM elements. Morphologically abnormal lymphocytes may also be recognized on occasion, but lymphomas are better identified in the BM biopsy.

Iron staining should be done on the BM aspirate rather that the biopsy because the decalcification process can artifactually give a negative result. If an aspirate specimen is not available, then the iron stain may be attempted on the biopsy and is meaningful when positive but not necessarily when negative. Iron stain findings are described simply: there may be no visible iron, adequate storage iron, or too much iron. Formal criteria are not used. The absence of iron in the BM aspirate has been said to be the gold standard for the diagnosis of iron deficiency, but this is questionable. The presence of storage iron reliably rules out iron deficiency. Excessive BM iron is seen with hereditary hemochromatosis and iron overload resulting from repeated blood transfusions. **Ring sideroblasts** (erythroid cells with iron granules encompassing two-thirds of the nuclear circumference) occur with MDS and other conditions such as heavy metal toxicity. The presence of iron in macrophages but not in erythroid cells (siderocytes) is suggestive of ACD.

Bone marrow differential cell count: A 500 cell manual differential count should be performed by an experienced technician on all adequate BM aspirate specimens. The differential cell count should not be reported if spicules are absent. Normal ranges are not usually given in the BM report because the interpretation of deviations from these ranges must be done within the context of the totality of the BM findings.

Bone Marrow Biopsy

Several observations are best made on the BM biopsy:

- Estimation of the BM cellularity
- Megakaryocyte adequacy
- Recognition of lymphoma
- Identification of metastatic carcinoma
- BM fibrosis
- Metabolic bone disease

In some cases, the biopsy is the only specimen available for examination.

Bone marrow cellularity is estimated by looking at the areas between the trabeculae and roughly assessing the percentage occupied by hematopoietic cells compared to fat. Cellularity is approximately 100% at birth and decreases by 1% per year. This valuation is usually stated in 10% increments as a reflection of its imprecision. The marrow just beneath the cortex is relatively hypocellular and does not yield an accurate estimate. If the specimen is small and subcortical, then this should be stated in the report and the estimation of cellularity omitted.

The myeloid to erythroid ratio (M:E) has traditionally been estimated on the BM biopsy as a cross-check of that calculated from the aspirate differential cell count, because of the dogma that the gathering of normoblasts into "erythroid islands" makes their distribution on the aspirate smears uneven. This is usually unnecessary if a 500 cell differential has been done on an adequate specimen. When comparing the estimate made on the biopsy to the M:E calculated from the 500 cell differential count of the aspirate, the latter usually results in a slightly lower ratio. This is probably because it is more difficult to morphologically differentiate early erythroid precursor cells from early granulocytic cells in the biopsy resulting in undercounting of the erythroids. The normal range for the M:E ratio is 1.2:1 to 5:1.

Myeloid and erythroid maturation: Delineation of the myeloid and erythroid cells into specific stages of maturation is somewhat more problematic in the biopsy compared to the aspirate due to crowding of the cells, because the hematoxylin and eosin (H and E) stain is less well suited for highlighting features such as cytoplasmic granules and because the biopsy is examined using less magnification. For these reasons, the range of maturation is described in general terms and a formal differential cell count is not done. Range of maturation of each cell line can be determined with practice. The imprecise process of differentiating early stage myeloid from erythroid cells, and blasts from slightly more mature cells (promyelocytes and myelocytes) can be aided by IHC stains.

Megakaryocytes: Estimation of adequacy is done by examining the biopsy on 100x (10x lens, 10x ocular) magnification and identifying one to three megakaryocytes per field. Morphologic abnormalities of megakaryocytes can be appreciated on the biopsy but not as well as in the aspirate.

Lymphocytes: The presence of lymphoid aggregates in BM biopsy sections is frequently discernable from low power. The differentiation of benign lymphoid aggregates from lymphoma may require special studies such as FCM or IHC. Some lymphomas tend to occur in an interstitial pattern and are not identifiable without IHC. The classic example of this is ALCL. The morphology of lymphoma in the BM fails to correlate with that seen in the lymph node in a significant number of cases. Involvement of the BM by lymphoma is discussed fully in a separate chapter.

Extrinsic cells: Involvement of the bone marrow by metastatic carcinoma is usually readily evident. Carcinoma cells differ from hematopoietic and lymphoid cells by their greater size, more severe atypia and cohesiveness. Metastatic carcinomas are commonly accompanied by a fibrotic tumor stroma. IHC plays a crucial role in determining the site of the primary tumor.

Touch Preparations (TP) of the BM Biopsy

Because the morphology is comparable to that of an aspirate, Wright-Giemsa-stained TPs of the BM biopsy fragment can be extremely helpful in cases in which the aspirate specimen is inadequate. TPs should be done any time the person collecting the specimen feels there has been a dry tap or in conditions that are associated with BM fibrosis.

Clot Sections

Clot sections are prepared by clotting material from one of the aspirate tubes with thrombin and submitting the clot for tissue processing. The morphologic information rendered is equivalent to the biopsy. The importance of clot sections is as an archival source of material for FISH and/or PCR studies which cannot be done on decalcified biopsy sections or on aspirate tubes that are more than a few days old.

The Future of Bone Marrow Reporting: Synoptic Reports

A synoptic BM report takes the data from the traditional narrative report and presents it in list or tabular form. Synoptic reports are superior to traditional BM reports in the following ways:

- Large amounts of complex information are more easily assessable by the clinician.
- Important data is less likely to be omitted.
- Reduced dictation time
- Reduced transcription time
- Reduced transcription errors
- Facilitates the integration of results from other studies (FCM, FISH, cytogenetics, etc.)

The primary disadvantages to synoptic reports are that they are physically long when printed and may contain much information that is irrelevant to a particular case.

Synoptic reports are currently not in wide routine use. The College of American Pathologists recommends and provides an outline for the synoptic reporting of hematologic neoplasms but provides no specific guidelines for non-neoplastic cases. The British Royal College of Pathologists has developed a minimum dataset (their name for a synoptic report) for lymphoma in the BM, but datasets for other diagnoses are pending.

There are a multitude of forms in which a synoptic report might appear. Since this practice is in its infancy, the pathologist has a great deal of freedom to experiment and discover what works best in his practice setting. This is how the previous example of a normal BM report might look in synoptic form.

Sample Synoptic Bone Marrow Report

Clinician: John Smith, M.D.
Clinical History: Staging of follicular lymphoma
Peripheral Blood Diagnosis: Normal peripheral blood smear.
Bone Marrow Diagnosis: Negative for lymphoma;
trilineage hematopoiesis
Diagnostic Comment: None

Peripheral Blood Description

CBC:

WBC	6.1 K/mm3	PLT	245.0 K/mm3
RBC	4.3 M/mm3	MPV	9.8 fl
HGB	14.3 g/dl	Seg	68.0 %
HCT	40.6 %	Band	2.0 %
MCV	89.7 fL	Ly	16.0 %
MCH	31.1 pg	Mono	10.0 %
MCHC	34.7 g/dl	Eos	4.0 %
RDW	12.0 %	Baso	0.0 %

Red Blood Cells:
Morphology normal
No agglutination/rouleaux
White Blood Cells:
Segmented neutrophils morphologically normal
Lymphocytes, small and mature in appearance
Monocytes normal in number; morphologically unremarkable
Eosinophils normal in number; morphologically unremarkable
Basophils not identified morphologically
Platelets:
Estimated number normal
Morphology normal

Bone Marrow Aspirate Description

Cellularity normal; adequate for evaluation
Myeloid Cells: full range of maturation present, morphology normal
Erythroid Cells: full range of maturation present, morphology normal
Megakaryocytes: adequate in number; morphology normal
Plasma Cells and lymphocytes not increased
Stainable Iron: adequate, without ring sideroblasts
Differential Cell Count:

Seg	10 %	Meta	13 %
Band	13 %	Myelo	10 %
Mono	2 %	Pro	2 %
Ly	21 %	Blast	1 %
Eos	3 %	PCs	2 %
Baso	0 %	Other	0 %
Ery	20 %		

Bone Marrow Biopsy Description

Cellularity: 40%
Estimated M:E: 4:1
Erythroid and myeloid Cells: full range of maturation present
Megakaryocytes: normal
No evidence of leukemia, lymphoma or extrinsic cells
Special Studies: Flow cytometry negative for lymphoma

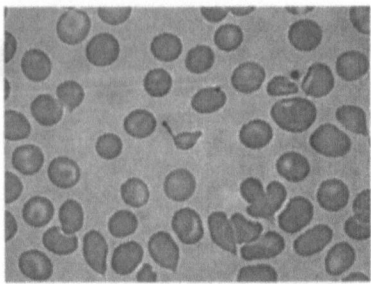

Figure 1.1. Acanthocyte (PB, 500x).

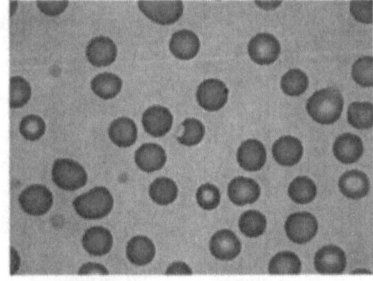

Figure 1.2. Bite cell (PB,500x).

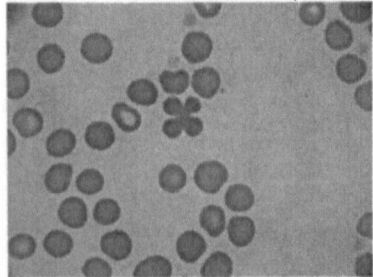

Figure 1.3. Bizarre poikilocyte (PB, 500x).

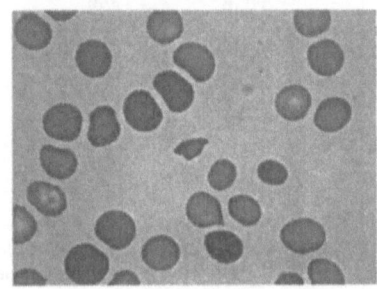

Figure 1.4. Schistocyte (PB, 500x).

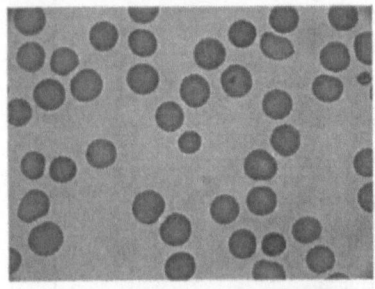

Figure 1.5. Spherocyte (PB, 500x).

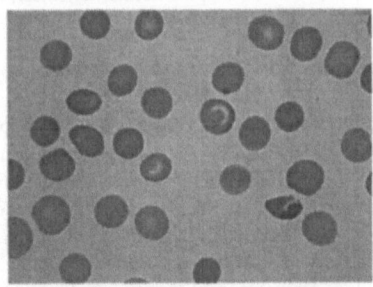

Figure 1.6. Stomatocytes (PB, 500x).

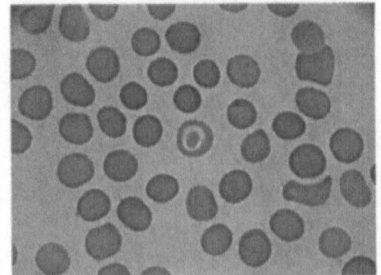

Figure 1.7. Target cell (PB, 500x).

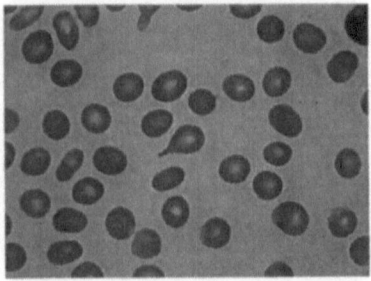

Figure 1.8. Teardrop cell (PB, 500x).

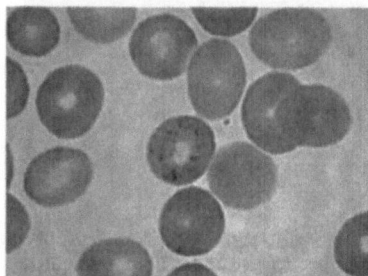

Figure 1.9. Basophilic stippling (PB, 1000x).

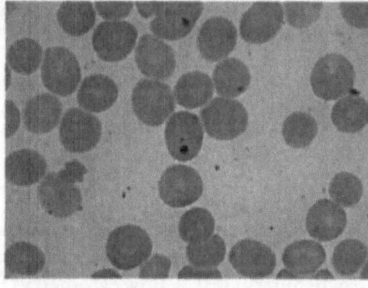

Figure 1.10. Howell-Jolly body (PB, 500x).

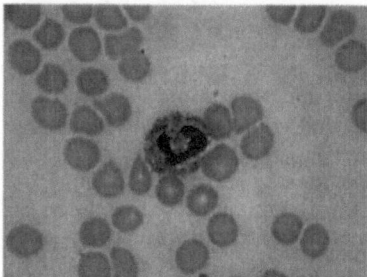

Figure 1.11. Neutrophil toxic granules (PB, 500x).

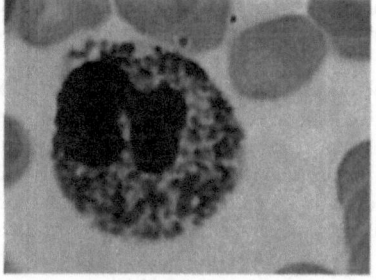

Figure 1.12. Toxic granules and Dohle bodies (PB, 1000x).

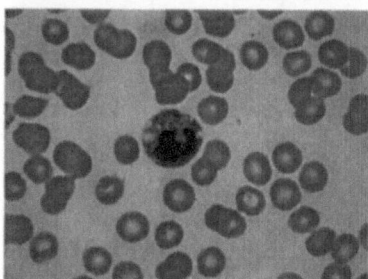

Figure 1.13. Neutrophil cytoplasmic vacuolization (PB, 500x).

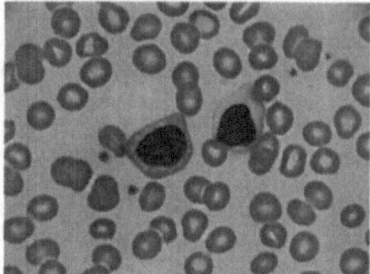

Figure 1.14. Downey cells (PB, 500x).

CHAPTER 2

Special Studies

Introduction

Great gains have been made in our knowledge of hematology based on immunophenotypic and genotypic data obtained by special studies such as immunohistochemistry (IHC), flow cytometry (FCM), fluorescence in situ hybridization (FISH), polymerase chain reaction (PCR) and others. The definitions of many entities in the new World Health Organization (WHO) classification of hematologic neoplasms are based on immunophenotypic and genotypic features, displacing morphology and clinical findings to secondary roles.

For the purpose of discussion, special studies used to evaluate BM and PB can be divided into those that yield immunophenotypic information and those that give genetic information. In general, the community hospital pathologist must have a deeper understanding of immunophenotyping because he is usually the lone interpreter of IHC and is commonly required to make important morphologic correlations with FCM data. Because of its importance in this setting, FCM is discussed fairly extensively but not to an extent to constitute an instructional text.

Genotyping is accomplished by a more varied group of methods that are rarely performed or interpreted in the community hospital setting. Integration of genetic findings into the diagnostic process primarily entails knowledge of specific associations, as opposed to the pattern recognition skills required to interpret immunophenotypic data. These associations are commonly listed in the interpretive report from the reference laboratory. Some genetic abnormalities are identified using immunophenotyping methods (e.g., IHC for the NPM-ALK gene in anaplastic large cell lymphoma and FCM to identify hypoploidy in acute lymphoblastic leukemia).

Enzyme cytochemistry is an older technique for determining cell lineage based on visualization of characteristic cytoplasmic chemicals by non-immunologic reactions and does not quite fit into either of the above categories. Cytochemistry has been largely supplanted by other tests but is still quite useful in some settings. Furthermore, cytochemistry is historically important as the basis for much of our original understanding of acute leukemias and for the French American British (FAB) classification system of hematologic neoplasms, which is still widely referenced. Cytochemistry is discussed in the chapters on acute leukemia.

Immunophenotyping

Normal and malignant cells express multiple proteins (antigens, markers) on their surfaces and within their cytoplasm that are characteristic of cell lineage or developmental stage. Monoclonal antibodies have been created to hundreds of these proteins and assigned cluster designations (CDs) (Table 2.1) as well as

Bone Marrow: A Practical Manual, by Daniel A. Cherry and Tomislav M. Jelic.
©2011 Landes Bioscience.

Table 2.1. Immunophenotypic markers of hematopoietic cells

CD1a	T-lymphoblasts, cortical thymocytes, Langerhans cells, and interdigitating dendritic cells
CD2	T cells, subset of NK cells, and malignant mast cells
CD3	T cells
CD4	Helper/inducer T cells, and monocytes/macrophages
CD5	T cells, MCL, and CLL
CD6	T cells and subset of B cells
CD7	T cells, NK cells, T-lymphoblasts, and 10% of AMLs
CD8	Suppressor/cytotoxic T cells and subset of NK-cells
CD9	Pre-B cells, monocytes, and platelets
CD10 (CALLA)	B-lymphoblasts, T-lymphoblasts, granulocytes, and epithelial cells of kidney
CD11a	Lymphocytes, granulocytes, and monocytes
CD11b	Myeloid cells and NK cells
CD11c	Hairy cells, monocytes/granulocytes, NK cells, and subset of T cells
CD13	Monocytes, granulocytes, myeloblasts, and promyelocytes
CD14	Monocytes, Langerhans cells, and weakly in maturing granulocytes
CD15	Reed-Sternberg and cells, granulocytes, monocytes, and epithelial carcinoma cells
CD16	NK cells, granulocytes, macrophages, mast cells, and fetal thymocytes
CD18	Leukocytes (platelets are CD18 negative)
CD19	B cells, myeloblasts in AML with t(8;21); plasma cells are negative
CD20	B cells; plasma cells are negative
CD21	Mature B cells and follicular dendritic cells
CD22	B cells; plasma cells are negative
CD23	Follicular dendritic cells, subset of T cells, monocytes, and CLL
CD24	B cells and granulocytes; plasma cells are negative
CD25	Hairy cells, activated T and B cells, and activated monocytes
CD26	NK cells, subset of T cells, thymocytes, and epithelial cells
CD27	Subsets of T cells, thymocytes, memory B cells, and NK cells
CD28	Thymocytes and EBV virus infected B cells
CD29a	Leukocytes and platelets
CD30	Anaplastic large cell lymphoma, Reed-Sternberg cells, activated B cells, and activated T cells
CD31	Endothelial cells, monocytes, platelets, granulocytes, and subset of lymphocytes
CD32	Monocytes, neutrophils, B cells, and platelets
CD33	Myeloid and monocytic precursors, and dendritic cells
CD34	Progenitor cells (blasts) and endothelial cells
CD35	B cells, subsets of T cells, monocytes, granulocytes, and erythrocytes
CD36	Platelets, monocytes/macrophages, and early erythroid cells
CD37	Mature B cells; pre-B and plasma cells are negative
CD38	NK-cells, B cells, subsets of T cells, and plasma cells
CD40	Mature B cells; plasma cells are negative
CD41	Platelets and megakaryocytes

continued on next page

Table 2.1. Continued

CD42a	Platelets and megakaryocytes
CD43	T cells, activated B cells, plasma cells, and myeloid cells
CD44	Hematopoietic and non-hematopoietic cells; platelets are negative
CD45	(leukocyte common antigen) all hematopoietic cells except erythroid cells, follicular dendritic cells, Langerhans cells, Reed-Sternberg cells, plasma cells, and sometimes anaplastic large cell lymphoma
CD45 RA	Most of the lymphocytes (B and T)
CD45 RO	Activated and memory T cells, subsets of B cells, and monocytes
CD52	B and T cells, monocytes, most lymphomas, and subset of myeloma
CD54	Activated B and T cells, macrophages, and other cells
CD55	Hematopoietic cells
CD56	NK-cells, subset of T cells, and neoplastic plasma cells
CD57	NK-cells and subsets of T cells
CD61	Platelets, megakaryocytes, and monocytes
CD62L	Most B cells, naïve T cells, monocytes, and NK-cells
CD64	Monocytes/macrophages and granulocytes
CD69	Activated lymphocytes
CD71	Proliferating cells
CD79a & CD79b	Pre-B and mature B cells; plasma cells are negative
CD99	Ewing sarcoma and precursor T and B cells
CD103	Intraepithelial lymphocytes, thymus, and spleen
CD117	Myeloblasts, promyelocytes, mast cells, and gastrointestinal stromal tumor
CD123	Myeloid precursors, mast cells, basophils, DC2 cells, and subsets of lymphocytes
CD138	Plasma cells and epithelial cells; mature B cells negative
CD235	Erythroid precursors
CD246/ALK	Overexpressed in anaplastic large cell lymphoma; positive in subset of normal central nervous system cells
HLA-DR	B cells, activated T cells, monocytes, and myeloblasts; negative in promyelocytes
TdT	B-lymphoblasts, T-lymphoblasts, hematogones, and thymocytes
Cyclin D1 (bcl-1)	Low-level expression in normal cells. Overexpressed in mantle cell lymphoma, in subset of HCL, and in subset of plasma cell myeloma
Bcl-2	Hematopoietic and nonhematopoietic cells; overexpressed in B-lymphomas
Ccl-6	Germinal center B cells
Annexin A1	HCL and granulocytes
DBA.44	HCL and subsets of many B-cell lymphomas
MUM1/Irf4	Non-germinal center B cells
BOB1	B cells and plasma cell myeloma
OCT2	B cells and plasma cell myeloma
PAX5/BSAP	B cells and AML with t(8;21); plasma cells negative
ZAP70	T cells and unmutated B-CLL
TCL-1	T-cell prolymphocytic leukemia and many B-cell neoplasms

names according to the targeted proteins. They are in standard use to visualize cellular antigen expression on glass slides using IHC and in fresh fluid specimens or suspensions of morsellated solid tissue specimens using FCM. An exception is antibodies to kappa and lambda, which are typically polyclonal because of antigenic heterogeneity.

Flow Cytometry

The **basic principles** of FCM are simple. Cells in suspension are flowed through an aperture within a liquid sheath and individually illuminated with lasers scattering the parallel light beams, which then strike detectors producing electrical impulses that are analyzed using sophisticated computer software and depicted with scattergrams or histograms. Bone marrow and PB are well suited for FCM, because they consist of cells suspended in liquid. If the cell is large (granulocyte, blast), it will stop more of the parallel rays than a small cell, and this optical characteristic is observed by the forward scatter (FSC) detector. Greater internal complexity of the cell (cytoplasmic granulation, nuclear irregularities) causes scattering of the laser light that is detected by the side scatter (SSC) detector.

Monoclonal antibodies, which selectively bind to proteins that are characteristic for cell lineage or developmental stage, can be labeled with a fluorochrome that causes the cell to emit a specific wavelength of light (fluoresce) when illuminated by the laser. An individual cell can be evaluated simultaneously for 4 or even 10 markers because the antibodies can be conjugated with multiple fluorochromes emitting unique, non-overlapping wavelengths of light when excited by various lasers. Filters are used to prevent light of other wavelengths from reaching the photodetectors.

Each particle (cell, platelet, contaminant, etc.) that is **interrogated** (illuminated) by the laser is referred to as an **event**, which can be graphically displayed. By convention, FSC data is depicted on the Y-axis using a linear scale such as 0, 200, 400, 600, 800 and 1000 (Table 2.2). Side scatter is plotted on the X-axis using a linear scale such as 0 to 1000, or 0 to 1024. Fluorescence intensity is usually reported in relative rather than absolute units, since recognition of abnormal cell populations is based on aberrant antigen expression compared to normal cells, not the absolute intensity of fluorescence. Relative intensities are reported semi-quantitatively in log scale (Table 2.3).

An initial CD45/SSC scattergram of the BM or peripheral blood (PB) is done to get an overview of the cell populations present in the sample. In this

Table 2.2. Example of linear scale of FSC data

Quantitative Scale	Descriptive	Corresponding Size	Example
<200	Debris	Very small	Platelets, platelet clumps, contaminants
200-400	Low FSC	Small	Normal small lymphocytes
400-600	Intermediate FSC	Medium	Monocytes
>600	High FSC	Large	Blasts

Table 2.3. Fluorescence intensity scale

First decalog	10^0-10^1	Negative fluorescence (-)	
Second decalog	10^1-10^2	Weak (or dim) intensity (+)	
Third decalog	10^2-10^3	Moderate intensity (++)	
Forth decalog	10^3-10^4	Strong (or bright) intensity (+++)	

way, normal and sometimes abnormal cell groups may be identified and **gated** by the flow cytometrist; that is, a line is drawn around them using computer software so that they can be examined using specific monoclonal antibodies. For example, blasts are commonly identified using CD45 and SSC. Panels of monoclonal antibodies are then applied to establish a detailed immunophenotype. The characteristic antigen profiles for hematologic cells are presented in Table 2.4.

Table 2.4. Characteristic antigen profiles of hematologic cells

Myeloblasts: CD34, CD117, CD33, CD13, and HLA-DR positive

Promyelocytes: CD117, CD13, CD33 positive; HLA-DR negative

Monoblasts: CD4, CD14, CD64 positive; CD34 positive or negative

Precursor B lymphoblasts: CD34, CD19, CD10, TdT positive; surface immunoglobulin light chain and CD20 negative

Pre-B lymphoblasts: CD19, CD10, cytoplasmic μ (mu heavy chain), and TdT positive; CD34 and surface immunoglobulin negative; CD20 positive or negative

B-lymphoblasts (B-cell acute lymphoblastic leukemia, Burkitt lymphoma): CD19, CD20, CD10, and monoclonal kappa or lambda positive; TdT negative

Precursor T lymphoblasts: CD7 and TdT positive; CD1, CD2, surface CD3, CD4, CD5 and CD8 negative

Early cortical lymphocytes: CD1a negative; CD2, cytoplasmic CD3, CD5, CD7 and TdT positive

Late cortical lymphocytes: CD1a, CD2, cytoplasmic CD3, CD4, CD5, CD7 CD8 and TdT positive

Medullary lymphocytes: CD1a negative; CD2, cytoplasmic and surface CD3, CD4 or CD8 positive; CD5, CD7 positive; TdT positive or negative

Follicular lymphoma cells (centrocytes, centroblasts): CD19, CD20, CD10, surface immunoglobulin light chain monoclonal kappa or lambda positive; TdT-negative

B-cell chronic lymphocytic leukemia/small lymphocytic lymphoma cells: CD19, CD23, and CD5 positive; CD20 dim positive; surface immunoglobulin light chain dim positive or negative; FMC7 negative

Mantle cell lymphoma: CD5, CD19, and FMC7 positive; CD20 and surface immunoglobulin light chain strongly positive; CD23-negative

Hairy cell leukemia: CD103, CD11c, CD25, CD19, CD20, and monoclonal kappa or lambda positive; CD5 and CD10 negative

Benign mast cells: CD117, CD33 positive; CD2 and CD25 negative

Malignant mast cells: CD117, CD33, CD2 and CD25 positive

The practical minimum instrument capability for the evaluation of hematology specimens is four colors, while more advanced flow cytometers have a 10-color capacity. This refers to the number of signals from different fluorochromes that can be detected on a single passage of a cell through the instrument, which, in turn, represents the number of antigens that can be examined in a single test tube and directly compared to one another. A typical BM aspirate or PB specimen is divided into three to five test tubes depending on adequacy. The total number of antigens that can be evaluated on the specimen is the number of colors the instrument is capable of analyzing multiplied by the number of tubes. Therefore, FCM instruments that can analyze more colors conserve specimen. The expression on a cell population of interest of antigens that are not analyzed in the same tube cannot be directly compared but, rather, relies on the skill of the flow cytometrist to recognize and gate the same cell group in each tube.

Antibody (marker) panels differ somewhat from one laboratory to another based on instrumentation, philosophy and personal preference. Philosophically, some laboratories determine their antibody panels in a stepwise manner, choosing the initial panel based on morphology or clinical suspicion and then applying additional rounds of antibodies if needed as the immunophenotype of the cells unfolds. Other laboratories use large comprehensive panels at the onset that are capable of immunophenotyping a wide variety of cells without first knowing what specific population is being evaluated. The former approach uses fewer reagents but requires more technologist time and is more economical for low volume laboratories, while the latter economically favors high volume laboratories. Furthermore, the pathologist interpreting the study at a large reference laboratory is less likely to have morphology and clinical data for correlation. The comprehensive approach is less likely to miss a small or unsuspected population of neoplastic cells. Whenever possible, the interpreter of the FCM data should correlate it with morphology.

Putting instrumentation and philosophy aside, every sample cannot be evaluated for every antigen. Specimen type serves as a good parameter upon which to base antibody panels. The principle is that some neoplasms are frequently encountered in certain body sites, and uncommonly in others. For example, hairy cell leukemia (HCL) is exceedingly rare in lymph nodes, as opposed to the spleen. A typical reference laboratory menu might include separate comprehensive panels for BM/PB and tissue/body fluids and additional small panels for specific diseases (myeloma, NK-cell tumors, ALL, etc.).

Applications for which FCM excels in the evaluation of BM and PB include the following:
- The diagnosis and classification of hematologic neoplasms by establishing monoclonality and characterizing the immunophenotype of the abnormal cells
- Assessment and quantification of residual disease after therapy, made possible by the ability of FCM to detect approximately one neoplastic cell among 10,000 normal cells
- Accurate enumeration of the absolute numbers and percentages of normal and abnormal cells, such as myeloblasts, lymphoblasts and non-Hodgkin lymphoma cells

- Determination of the density of cell surface antigens, which is useful in the diagnosis of mature B-lymphocytic neoplasms and essential for establishing the diagnosis of T-cell malignancies
- Assessment of biologic markers important for prognosis, including CD38 and ZAP-70 in B-cell chronic lymphocytic leukemia/small lymphocytic lymphoma (CLL/SLL), DNA content analysis (aneuploidy has prognostic significance in precursor B-lymphoblastic leukemia) and tumor proliferation rate by quantifying the number of cells in S phase (correlates with lymphoma grade)

Disadvantages of FCM include the following:

- Requirement of fresh tissue
- Cells stained with fluorochrome cannot be visualized for purpose of morphologic correlation.
- High cost and need for technical expertise
- Flow cytometry is not useful for the routine diagnosis of Hodgkin lymphoma because the malignant cells (Reed-Sternberg cells) are sparse and, although monoclonal, express neither surface nor intracytoplasmic immunoglobulin (sIg and cIg, respectively).

Protocols for collecting PB and BM for FCM may vary somewhat from one laboratory to another. In the author's hospital, PB and BM are collected in yellow top (ACD) tubes and kept at room temperature. If FCM is expected to be delayed, the tubes are refrigerated. Lavender top (EDTA) tubes are obtained for Wright-Giemsa smears and CBC, as well as a green top (heparin sodium) tube for chromosomal studies. FCM can be done on EDTA tubes, but cell viability is better maintained in ACD tubes.

Diagnostic Application of FCM

B-lymphocyte monoclonality is easily established using FCM to demonstrate abnormal (restricted) surface expression of either kappa or lambda Ig light chains, but not both. The normal range for the κ to λ ratio is approximately 1.0 to 2.5. This range is not strict, but rather a loose guideline that must be applied within the total context of the immunophenotype. Some clonal B cells demonstrate no surface Ig. **Plasma cells** express Ig light chains in their cytoplasm; therefore, the added step of cell membrane permeabilization must be performed in order to demonstrate clonality. **T-lymphocyte monoclonality** determination by FCM is less straightforward and involves the aberrant loss of pan-T-cell markers (usually CD7 or CD5, or loss of both CD4 and CD8), abnormalities of the CD4 to CD8 ratio (normal is approximately 0.9 to 2.5) co-expression of both CD4 and CD8 or changes in surface T-cell antigen densities. Newer strategies utilize panels of multiple antibodies to αβ T-cell receptor (TCR) gene antigens from which clonal gene rearrangement can be inferred.

B-cell CLL/SLL is the most common hematologic malignancy worldwide and has a characteristic immunophenotype that enables the identification of fewer than 1000 malignant lymphocytes in 1 μL of sample. Any adult with unexplained absolute lymphocytosis is a potential candidate for PB FCM in order to detect early disease. Malignant B cells of CLL/SLL express CD5 (a T-cell marker) and the pan B-cell markers CD19, CD20 and CD23 with monoclonal expression of sIg light chains. The CD20 and sIg light chain expression are dim, and sIg light chains frequently are not expressed at all. They are negative for FMC-7. This is in contrast to mantle cell lymphoma (MCL) cells, which strongly express CD20 and sIg light chains as well as FMC-7 but do not express CD23.

A subset of CLL/SLL is composed of naïve cells with unmutated Ig genes and is more aggressive than cases with mutated Ig genes. Some clinicians give these patients earlier therapy. The mutational status can be determined by DNA sequencing of the Ig heavy chain gene, but this procedure is laborious. Expression of CD38 and ZAP-70 by FCM correlate strongly, although imperfectly, with unmutated Ig gene status and, thus, are convenient surrogate markers for a worse prognosis. Flow cytometric measurement of CD38 is easy and reliable, but that of ZAP-70 is currently problematic and non-standardized.

Mantle cell lymphoma expresses CD5, CD19, FMC-7, CD20 (strong) and sIg light chains (strong) but is negative for CD23. Chromosomal studies showing t(11;14) or detection of bcl-1 (cyclin D1), either by IHC or FISH, provide confirmation of the diagnosis.

Follicular lymphoma (FL) characteristically expresses CD10 together with the pan B-cell markers CD19 and CD20. Immunoglobulin light chain expression is bright.

Burkitt lymphoma has a similar immunophenotype, the difference being the absence of cytoplasmic bcl-2 as opposed to over expression in FL. The presence of bcl-2 can be easily assessed by IHC on paraffin embedded tissue. The FCM technique is infrequently used, since fixation and permeabilization of the cell membrane are required.

The cells of **HCL** are seen in the lymphocytic/monocytic area of the FSC/SSC dot plot and characteristically express CD103, CD11c and CD25 as well as the pan B-cell markers CD19, CD20 and CD22. They strongly express monoclonal sIg light chains. Hairy cell leukemia is effectively treated with the purine analogues 2'-deoxycoformycin and 2-chlorodeoxyadenosine. The distinction of HCL from **HCL variants**, which do not respond to this therapy, is of obvious importance. Hairy cell variants express CD103 and CD11c but not CD25, and are usually negative for tartrate-resistant acid phosphatase (TRAP). Promising new markers for HCL include annexin A1 and CD123. CD103, CD11c and CD25 expression are not totally specific for HCL and may occur in other types of B-cell lymphoma but are usually less bright. For example, in **lymphoproliferative disorder, not otherwise specified, with lymphoplasmacytic morphology**, expression of CD11c and CD103 is heterogeneous, varying from negative to positive (trail pattern), while in HCL their expression is homogeneous so that the malignant cells on the scattergram form a circle, not a trail. **Splenic marginal zone lymphoma with villous lymphocytes (SMZLVL)** can be positive for CD11c, CD25 and TRAP but is negative for CD103.

Nodal MZL and extranodal MZL, including those of mucosa-associated lymphoid tissue (MALT lymphomas), do not have specific markers. They are monoclonal and express the pan B-cell markers CD19 and CD20. They can be differentiated from other morphologically similar lymphomas by the absence of CD5, CD10 and CD23. CD43 may or may not be expressed in MALT lymphomas and in many cases of nodal marginal zone B-cell lymphomas.

Table 2.5. Immunophenotype of frequently encountered B-cell neoplasms

	CD20	CD5	CD23	FMC-7	CD10	CD103	CD11c	CD25	bcl-1	bcl-2
CLL/SLL	+	+	+	-	-	-	-	-	-	+
MCL	+	+	-	+	-	-	-	-	+	+
FL	+	-	+	+	+	-	-	-	-	+
HCL	+	-	+	+	-	+	+	+	-	+

Careful scrutiny of Table 2.5 reveals a handful of antigens (CD5, CD10, CD19 and CD23) that are extremely helpful for differentiating the most common lymphomas (CLL/SLL, FL and MCL) from one another. Dual parameter scattergrams can be used to link these antigens to each other or to light chain expression so that the characteristic immunophenotypes of these neoplasms are recognized in a few glances.

Precursor B-lymphoblastic leukemia (B-ALL) is the most frequent malignancy in children. Cardinal immunophenotypic features are the expression of CD19, CD10 and cytoplasmic terminal deoxynucleotidyl transferase (TdT). Clinically relevant is the immunophenotypic differentiation into three subtypes: B-precursor ALL, pre-B ALL and B-ALL/Burkitt lymphoma (BL). These represent malignant counterparts of developmental phases of B-lymphocytes from progenitor to mature B cells and can be differentiated by using a small number of antibodies (Table 2.6).

B-precursor ALL (CALLA positive Pre-Pre ALL) and **pre-B ALL** correspond to the FAB classification of either L1 or L2, and BL to that of L3. CD34 is present in B-precursor ALL, cytoplasmic mu (mu$_c$) heavy chain in the pre-B ALL and sIg light chains (monoclonal, either kappa or lambda) in BL.

Hematogones are normal B-cell precursors and demonstrate a continuum of B-cell development from early progenitors to mature B-lymphocytes. Morphologically they resemble small lymphoblasts (L1 type) with high nuclear to cytoplasmic ratios, homogeneous chromatin and absent or inconspicuous nucleoli. The immunophenotype of hematogones is similar to that of lymphoblasts (positive for CD19, TdT and CD10). Hematogones are normally present in the bone marrow of children and can be seen in patients of any age who have PB cytopenias or regenerating BM after cytostatic therapy. They may be difficult to differentiate from residual lymphoblasts. Immunophenotypically, hematogones acquire the pan B-cell marker CD20 gradually, so early precursors do not express it at all and

Table 2.6. Immunophenotype of B lymphoblastic leukemias

	CD19	CD20	CD34	TdT	mu$_c$	sIg
B-precursor ALL	+	-	+	+	-	-
Pre-B ALL	+	-+	-	+-	+	-
B-cell ALL/BL	+	+	-	-	+	+

mu$_c$: cytoplasmic mu

mature B cells show bright CD20 expression. CD10 and TdT expression occur in the reverse pattern. Both of these characteristics are clearly displayed on the CD20/CD10 scattergram, where they form the shape of the letter U or J. Surface Ig light chains are expressed by a smaller fraction of hematogones and are polyclonal and dim. In contrast, B-lymphoblasts do not express sIg light chains, their expression of CD10 and TdT is homogeneous (forms circle rather than trail on scattergram), and they either do not express CD20 or, if they do, express it homogeneously (circle on CD20/CD10 scattergram, not trail).

Precursor T-lymphoblastic leukemia/lymphoma (T-ALL/LBL) comprises about 15% of ALLs in children and usually presents with a large mediastinal mass accompanied by peripheral lymphadenopathy and circulating lymphoblasts in the PB. The malignant cells are counterparts of the developmental phase of T cells from progenitor to late (medullary) thymocyte. However, the immunophenotype of T-lymphoblasts may not correlate with the known stages of T-cell development and, thus, may not exactly conform to a maturation arrest model. Generally speaking, T-cell neoplasms show either the loss or aberrant expression (dimmer or brighter than in normal counterparts) of T-cell markers. Common to all subtypes of T-ALL/LBL is the expression of CD7, cytoplasmic CD3 and TdT. For the purpose of contemporary therapy, it is not necessary to divide T-ALL/LBLs into multiple subtypes corresponding to the development of thymocytes but rather only into the pre-T and T-cell types. The malignant cells of pre-T-ALL/LBL express only CD7, cytoplasmic CD3 and TdT. T-cell-ALL/LBLs express these markers and additional T-cell antigens including CD1a, CD2, CD5 and both CD4 and CD8.

Large granular lymphocytes (LGLs) are either cytotoxic CD8$^+$ T cells or natural killer (NK) cells and normally comprise 10 to 15% of lymphocytes in the PB with an absolute number of 200 to 400 LGLs/μL. Reactive benign large granular cell lymphocytosis is usually caused by a viral infection (EBV, CMV, etc.). If increased LGLs persist in the PB for more than six months without an obvious cause, or LGLs comprise 40% or more of the WBCs (absolute number >2000 LGL/μL), then malignancy should be suspected. **Chronic T-large granular lymphocytic (T-LGL) leukemia** is a proliferation of malignant cytotoxic CD8$^+$ T cells usually accompanied by neutropenia, anemia, splenomegaly and BM involvement (features of Felty's syndrome). The normal immunophenotype of cytotoxic CD8$^+$ T cells is CD2, CD3, CD8, CD16 and T-cell receptor gene positive, and CD56 negative. Malignant CD8$^+$ T cells may exhibit down-regulation (dim) or up-regulation (bright) of some of the T-cell markers. Neoplastic proliferations of NK-LGLs are classified as **indolent natural killer large granular lymphocyte leukemia** in the WHO scheme. Natural killer cells express CD8, CD16 and CD56, and are negative for CD3 and the T-cell receptor gene.

FCM is an adjunct, rather than primary, tool in forming the diagnosis of **myelodysplastic syndromes (MDS)**. The ability of FCM to accurately determine the percentage of myeloblasts in the BM and PB is helpful for the sub-classification of MDS, and increased myeloblasts (left-shift) is the most specific FCM finding for MDS. Other features of myelodysplasia, including hypogranularity (decreased SSC) and distorted antigenic maturation of myeloid cells, can be detected by FCM; however, the use of these findings for diagnostic purposes is not widespread because they are difficult or impossible to quantify and are not standardized. Distortion of myeloid maturation is best established by observing the expression of CD13/ CD16, CD11b/CD13 and CD11b/CD16 in gated granulocytes. Another useful parameter is CD10 expression, which is lost or significantly decreased in dysplastic granulocytes. The specific use of FCM in the WHO classification system of MDS is discussed in the MDS chapter.

The main role of FCM in the realm of leukemias is to distinguish AML from ALL. The **cardinal myeloid markers** are CD13 and CD33. The **cardinal monocytic markers** are CD14, CD4 and CD64. CD117 is a marker of myeloblasts, promyelocytes and mast cells. Acute myeloid leukemia presents on the 45/SSC dot plot as an expansion of cells in the blast gate and decreased to absent granulocytes because of decreased myeloid maturation. The malignant cells of AML express all or some combination of the markers CD34 (hematopoietic progenitor cell marker), CD117, CD13 and CD33. If there is a monocytic component, the malignant cells express CD4, CD14 and/or CD64.

Acute promyelocytic leukemia (APL) cells exhibit CD13, CD33 and CD117 and are negative for HLA-DR, integrins (CD11a, CD11b, CD11c, and CD18) and sometimes CD34. HLA-DR negativity in APL is particularly valuable because it is positive in other AMLs. The differentiation from other types of AML is extremely important, because APL is treated differently and the administration of standard AML therapy can be rapidly fatal due to disseminated intravascular coagulation (DIC). FCM plays a crucial role in this regard because the hallmark of APL, the t(15;17) genetic abnormality, is detected by chromosomal and molecular studies (FISH, PCR) whose results may be delayed beyond the appropriate time in which therapy should be instituted. Another useful marker is CD2, which is positive in the microgranular variant of APL and may portend a poor prognosis.

In **acute megakaryoblastic leukemia**, the malignant cells are positive for the platelet markers CD41 and CD61 as well as the standard myelocytic markers. They are negative for CD42. In **acute erythroid leukemia**, there are large numbers of glycophorin positive cells, while myeloblasts (which need to comprise 20% or more of non-erythroid cells) are positive for some combination of the myelocytic markers CD33, CD13, CD117 and CD34 but typically lack CD45.

Acute myeloid leukemia, minimally differentiated, and AML, with maturation, may express CD7 and TdT. The T-cell antigen CD7 is not normally present on myeloid cells but appears as an early marker on leukemic myeloblasts. It may suggest a poor response to standard therapy. The B-cell marker CD19 in **AML, with maturation,** is highly associated with the t(8;21) abnormality. The expression of CD7 or CD19 does not indicate biphenotypic leukemia.

Biphenotypic leukemias express both myeloid and lymphoid markers on the same cell. Not all markers are equally specific for a particular lineage. Myeloperoxidase (MPO), cytoplasmic CD22 and cytoplasmic CD3 are the most specific markers for the myeloid, B-lymphocytic and T-lymphocytic lineages, respectively (surface CD22 and surface CD3 are also specific but may not be present in the early stages of maturation). If two of these are co-expressed by the leukemic cells, for example MPO and CD3, then the leukemia is biphenotypic. For other less specific markers, a scoring system is in use to establish whether they indicate true biphenotypic leukemia or just aberrant expression of other lineage antigens. Biphenotypic leukemia suggests that neoplastic transformation has occurred in the stem cell at a stage prior to lineage commitment and may portend a worse prognosis. In **biclonal acute leukemia** there are two distinct leukemic cell populations of different phenotypic lineages.

Immunohistochemistry

The **basic principle** of IHC is the exploitation of the specificity of the antigen-antibody relationship to identify tissues harboring targeted epitopes (antigens). The antibody is labeled with a fluorochrome or is enzyme bound (antibody-antigen binding triggers the enzyme, which acts on a substrate and produces color) in order to visualize the specific antigen. IHC has tremendously advanced pathology, enabling accurate classification of difficult to diagnose malignancies and poorly differentiated neoplasms that are impossible to differentiate by standard hematoxylin and eosin staining. Malignant cells harbor many proteins; and some, as described above, are associated with cell type, lineage, maturation stage, etc. Antibodies have been synthesized to these proteins (antigens) that selectively bind to their target antigens.

In the **direct method**, the antibody is labeled with a fluorochrome, colloidal gold, enzyme or biotin that will make the antigen visible when binding occurs. The **indirect (sandwich) method** uses two layers of antibodies in order to enhance the signal and, thus, increase sensitivity without decreasing specificity. The primary antibody that binds to its target antigen in the tissue is not labeled. The binding is visualized by a secondary antibody conjugated with an enzyme (usually peroxidase) that will react with the substrate and produce color on binding, indicating and locating (cytoplasm or nucleus) the presence of the antigen. The secondary antibody is an anti-IgG antibody against the Fc and Fab portions of the primary antibody. This means that two secondary antibodies, one on the Fc epitope and the other on the Fab epitope, bind to every bound primary antibody, doubling the staining intensity in comparison with the direct method. The use of a tertiary antibody amplifies the staining even more.

The three great advantages of immunohistochemistry are the following:

1. Malignant cells stained for a particular antigen are visible by light microscope.
2. Numerous useful specific antibodies have been synthesized (Table 2.7).
3. Paraffin embedded tissue fixed in standard fixatives (formalin, B5, Bouin's solution, Zenker's fluid) is used and, thus, obviates need for fresh tissue and enables use of archived material.

The usefulness of IHC for the diagnosis of neoplasms in the BM is evident throughout this manual.

Genotyping

Hundreds of genetic defects have been associated with hematologic neoplasms, some highly specific and others less so. Identification of these abnormalities serves many purposes in hematopathology:

- Some are associated so closely with a particular neoplasm so as to define or be diagnostic of that neoplasm (e.g., PML/RARα in APL).
- Less specific associations, such as the non-random occurrence of certain abnormalities with particular neoplasms (e.g., trisomy 3 in MALT lymphoma) also have diagnostic utility.

Table 2.7. IHC antibodies useful in routine BM evaluation

ALK-1 Anaplastic large cell lymphoma marker

Bcl-2 Oncogene marker in follicular lymphoma

Bcl-6 Follicular center cell marker

CD117 Myeloblast marker, mast cell marker

CD20 B-cell marker

CD3 T-cell marker

CD30 Reed-Sternberg cell marker, anaplastic large cell lymphoma marker

CD34 Stem/blast cell marker

CD4 Macrophage marker and helper/inducer CD4⁺ T-cell marker

CD61 Megakaryocytes, platelets, acute megakaryoblastic leukemia

CD68 Macrophage marker

CD163 Monocyte marker

CD79a B-cell marker (especially useful if the patient was treated with the anti-CD20 drug rituximab)

CD8 Suppressor cytotoxic CD8⁺ T-cell marker

CD99 Ewing sarcoma cell marker

Chromogranin A neuroendocrine cell marker, neuroblastoma cell marker

Cytokeratin cocktail (AE1/AE3) highlights metastatic carcinoma cells in the marrow

DBA-44 Hairy cell leukemia marker

Estrogen Receptor Marker of breast origin for metastatic carcinoma

Factor 8 Megakaryocytes, platelets, acute megakaryoblastic leukemia

GCDFP-15 (gross cystic disease fluid protein) breast carcinoma marker

MART-1 melanoma cell marker

PLAP (placental alkaline phosphatase) seminoma cell marker

PSA (prostatic specific antigen) highlights marrow metastases from prostate cancer

Synaptophysin neuroendocrine cell maker, neuroblastoma cell marker

Thyroglobulin thyroid gland tumors

TRAP Tartrate-resistant acid phosphatase hairy cell leukemia marker

- Establishment of monoclonality of a B-cell or T-cell population
- Differentiation of neoplastic from reactive cellular proliferations (e.g., chronic myelogenous leukemia vs. leukemoid reaction)
- Separation of neoplastic (MDS) from non-neoplastic (nutritional or physiologic) etiologies for cytopenias
- Many abnormalities have established prognostic implications (e.g., hyperploidy in ALL).
- A growing number of abnormalities have targeted therapies (e.g., BCR/ABL and C-KIT).

The most commonly used techniques for genotyping PB and BM are cytogenetics (CG), FISH and PCR. Southern blot is occasionally used for Ig and T-cell receptor gene rearrangement studies. Microarray testing is a highly promising new technique that is certain to have wide applications in the near future.

Cytogenetics

The **basic principles** of CG could not be simpler. Metaphase chromosomes are directly examined for structural abnormalities. PB and BM are submitted for CG in sodium heparin (green top) tubes under sterile conditions to prevent bacterial and fungal overgrowth, and are expeditiously processed to maximize viability. The cells are cultured for about 24 to 48 hours, inhibited in metaphase using colcemid, and collected. Cells that are normally incapable of replication, such as mature lymphocytes, must be stimulated to do so with mitogens such as phytohemagglutinin. When maximum growth is achieved, the cells are swollen hypotonically and dropped on a slide, rupturing the cell membranes and leaving the chromosomes slightly separated from each other. The chromosomes are then stained and digitally photographed. A trained technician using computer software then arranges them into a karyogram that can be examined for abnormalities. Twenty cells are examined, since some may demonstrate culture artifacts.

The major **advantage** of CG over FISH and PCR is that the entire genome is examined rather than using probes for specific abnormalities. This is analogous to asking an open-ended question rather than a yes or no question. In other words, you don't necessarily have to know what you're looking for ahead of time.

The **disadvantages** of CG include the following:

- Inability to detect small abnormalities
- The requirement for fresh viable cells that are capable of replication when stimulated
- Slow turnaround time (TAT) of about five days
- High level of technical expertise required for analysis

Fluorescence In Situ Hybridization

Basic principles: In hybridization assays, a probe consisting of a fluorescently labeled known fragment of nucleic acid is used to interrogate unknown fragments of nucleic acid for a structure that it is able to complementarily form matching base pairs. In situ refers to the fact that the hybridization assay is performed on an intact cell. To perform FISH, PB or BM collected in a green (sodium heparin) or lavender (EDTA) top tube, or a BM clot section embedded in paraffin are placed on slides, the cellular DNA is denatured and a fluorescently labeled single stranded nucleic acid probe is placed on the slide under conditions that are favorable to hybridization. The non-hybridized DNA and probe are then washed away, and the slide is examined using a fluorescent microscope or by computer image analysis.

Advantages of FISH include the following:

- Frozen, fresh, fixed or paraffin-embedded (clot section) tissue can be used.
- Allows morphologic correlation, since the results are analyzed in intact cells
- Rapid TAT with results available in as little as a few hours
- Technically simple, relative to other molecular methods

- Probes are available for most clinically significant abnormalities.
- Excels at detecting abnormalities spread over large areas of the gene

Disadvantages of FISH include the following:

- Must know what genetic defect you are looking for in order to choose appropriate probes
- Mediocre analytic sensitivity
- Cannot detect small (sub-megabase) changes in DNA
- Cannot be done on decalcified BM biopsy specimens
- Relatively expensive

Polymerase Chain Reaction

Basic principles: PCR is a process by which very large quantities of a targeted nucleic acid sequence can be synthesized, facilitating detection or analysis of the product. PB or BM collected in a green (sodium heparin) or lavender (EDTA) top tube is placed in the PCR mixture, which is composed of DNA polymerase, two oligonucleotide primers, deoxynucleotide triphosphates (dATP, dCTP, dGTP, dTTP), $MgCl_2$, KCl and Tris-HCl buffer. This mixture is heated, causing **denaturation** of the target strands of DNA, and then cooled to allow **annealing** of the primers to the target DNA. The DNA polymerase initiates **extension** of the primer at the 3' end. Heating then disassociates the extension products from the target DNA, which serve as templates for additional rounds of annealing and extension. By this process, subsequent cycling can yield millions or billions of copies of the target DNA sequence. These steps are performed in a **cycler** under programmable computer control. The primers are usually labeled with a fluorescent tag, so the PCR product can be detected. In some cases, the product is analyzed by capillary gel electrophoresis or other methods.

There are many **variations** on PCR that allow the amplification of RNA targets (reverse transcriptase PCR or RT-PCR), increase the sensitivity and specificity (nested PCR), allow the amplification of two or more probes in the same reaction mixture (multiplex PCR) and allow the simultaneous amplification and detection of the target sequence (real-time [homogeneous, kinetic] PCR).

Advantages of PCR:

- Good analytic sensitivity, especially for evaluation of minimal residual disease (MRD)
- Good TAT that is only slightly longer than that for FISH
- Excels at detecting small gene abnormalities, including point mutations

A **disadvantage** of PCR is that it is poor at detecting abnormalities spread over large areas of the gene, because of the requirement for small (oligonucleotide) probes. This applies to the abnormalities associated with many lymphomas.

Southern Blot

In Southern blot, DNA in the test sample is enzymatically digested, the DNA fragments are separated using agarose gel electrophoresis, transferred to a nylon membrane and hybridized with chemically labeled probes. The bands are then analyzed by chemiluminescent techniques.

Clinical Applications of Genotyping

Hematologic neoplasms that are defined by genetic abnormalities are listed in Table 2.8 along with a few brief comments. In this table it can be seen that diseases from all of the major categories are represented.

For BMs done to evaluate **myeloproliferative** symptoms (usually elevated cell counts on the CBC), CG and molecular testing for BCR/ABL and JAK2 should be done in all cases. The definition of CML requires the presence of BCR/ABL, and the definitions of the other myeloproliferative neoplasms require its absence. Reactive causes of elevated cell counts are not associated with JAK2. Because JAK2 is positive in virtually all cases of polycythemia vera, this abnormality has become an important element of its diagnosis. Testing for BCR/ABL should be done using FISH since it is cryptic by CG in 10% of cases. JAK2 is identified by PCR. Cytogenetics is also done in these cases to rule out unsuspected abnormalities and because additional defects have prognostic significance for CML. If there is a substantial eosinophilia, then PDGFRA, PDGFRB and FGFR1 status should ascertained.

Cytogenetics plays an important role in the evaluation of BMs done for **cytopenias** because the presence of abnormalities is highly useful for confirmation of

Table 2.8. Hematologic neoplasms defined by genetic abnormalities

Disease	Gene or Synonym of Gene	Method of Detection	Comment
CML	BCR/ABL1; Philadelphia chromosome	FISH	10% cryptic by CG
Polycythemia vera	JAK2	PCR	Present in 95% of cases and rest have exon 12 variant
Systemic mastocytosis	KIT mutation	PCR, IHC	Seen in 95% of cases if sensitive methods used
Myeloid and lymphoid neoplasms with PDGFRA rearrangements	FIP1L1-PDGFRA; CHIC2 gene deleted	RT-PCR (FIP1L1); FISH for CHIC2 deletion	Most are cryptic; sometimes nested RT-PCR required
Myeloid neoplasms with PDGFRB rearrangements	ETV6-PDGFRB	RT-PCR, CG	Most have CG+ for t(5;12) (q31~33;p12) but 17 variants of gene; RT-PCR with probes to all is preferred
Myeloid and lymphoid neoplasms with FGFR1 abnormalities	FGFR1	CG, FISH	Eight different translocations with 8p11 breakpoint; trisomy may be seen by CG as secondary abnormality

continued on next page

Table 2.8. Continued

Disease	Gene or Synonym of Gene	Method of Detection	Comment
MDS with isolated del(5q)	Del(5q)	CG, FISH	Rare subset also have JAK2; if other abnormalities present, then this diagnosis is excluded
AML with t(8;21) (q22;q22)	RUNX1-RUNX1T1	CG, FISH	70% have additional genetic abnormalities
AML with inv(16) (p13.1; q22) or t(16;16)(p13.1; q22)	CBFB-MYH11	FISH, RT-PCR, CG	Subtle on CG, so RT-PCR or FISH preferred; 40% have additional gene abnormalities; rare cases of AML and CML with this abnormality and BCR/ABL, usually associated with disease progression
APL with t(15;17) (q22;q12)	PML-RARA	FISH, CG, PCR	40% have additional genetic abnormalities including FLT3 mutations; some AMLs have variant RARA translocations, only some of which are sensitive to ATRA therapy
AML with t(9;11) (q22;q23)	MLLT3-MLL	FISH, CG	Secondary CG abnormalities are common but do not affect survival
AML with t(6;9) (p23;q34)	DEK-NUP214	FISH, CG	Sole abnormality in most cases, but some have FLT3 mutations
AML with inv(3) (q21;q26.2) or t(3;3) (q21;q26.2)	RPN1-EVI1	FISH, CG	Additional abnormalities common; cases of CML may acquire these abnormalities associated with disease progression
Acute megakaryocytic leukemia with t(1;22) (p13;p13)	RBM15-MLK1	FISH, CG	Usually is sole abnormality
AML with mutated NPM1	Nucleophosmi-1 gene	PCR, IHC	On IHC demonstrates cytoplasmic rather than nuclear positivity
AML with mutated CEBPA	NA	PCR	Cryptic by CG; FLT3-ITD mutations in about one-third

continued on next page

Table 2.8. Continued

Disease	Gene or Synonym of Gene	Method of Detection	Comment
B-ALL/LBL with t(v;11q23)	MLL rearrangement	FISH	Cases with 11q23 deletion excluded from this diagnosis
B-ALL/LBL with t(12;21) (p13;q22)	TEL-AML (ETV6-RUNX1)	FISH	Sometimes seen in children who develop leukemia years later
B-ALL/LBL with hyperdiploidy	>50 and usually <60 chromosomes	FCM, CG	Usually is sole abnormality
B-ALL/LBL with Hypoploidy	<46 chromosomes	FCM, CG, FISH	May be missed on CG due to re-duplication; other abnormalities may also be present
B-ALL/LBL with t(5;14) (q31;q32)	IL3-IGH	CG, FISH	Usually detected by CG
B-ALL/LBL t(1;19) (q23;p13.3)	E2A-PBX1 (TCF3-PBX1)	FISH, CG	Subset of hyperdiploidy cases have this abnormality, but not the characteristic phenotype, and are excluded from this diagnosis
MCL	t(11;14) (q13;q32), cyclin D1, PRAD1	IHC, FISH	FISH with broad coverage DNA probes preferable due to heterogeneity of the gene
ALK positive large B-cell lymphoma	t(2;17) (p23;q23), CLTC-ALK	FISH, CG	Few cases have t(2;5) (p23;q35), as do the T-cell anaplastic large cell lymphoma cases
Burkitt lymphoma	t(8;14), C-MYC	FISH, CG	Several variants so FISH with broad range probes is preferable
Anaplastic large cell lymphoma, ALK positive	t(2;5) (p23;q35), NPM-ALK	IHC, FISH, RT-PCR	Several variants are not detectable by FISH or RT-PCR, so IHC is preferable; IHC positivity distribution within cell correlated with variant translocations

2

Table 2.9. important cryptic genetic abnormalities in hematology

Abnormality	Disease
t(12;21)	Pediatric ALL
BCR/ABL	10% of cases of CML
FLT3	CG normal AML
NPM1	CG normal AML
JAK2	MPNs, especially PV
CHIC2	Myeloid and lymphoid neoplasms with PDGBFRA rearrangements

the diagnosis of MDS. Since the genetic defects associated with MDS are almost always seen on CG, molecular testing is generally not needed in these cases.

The WHO classification of **AML** now includes nine entities associated with recurrent genetic abnormalities. Most of these are seen by CG. The most important application of molecular testing with AML is the use of FISH for the rapid identification of PML-RARA in suspected cases of APL. Whether or not AMLs with normal CG should be subjected to molecular testing for NPM1, CEBPA or FLT3 mutations should be guided by the needs of the clinical hematologist.

Most of the clinically significant genetic lesions associated with **B-ALL/ LBL** are now incorporated in the WHO classification. Of these, t(v;11q23) and hypoploidy should be looked for using FISH, since the former is not seen and the latter is usually missed by CG.

Some of the important genetic abnormalities that are cryptic by GC are listed in Table 2.9.

Lymphoma in the PB and BM is usually adequately evaluated by FCM without the need for genotyping. When genotyping is warranted, molecular techniques are preferable since most of these neoplasms consist of cells that will not enter metaphase without stimulation. Typically, a specific abnormality is being sought for which suitable FISH and/or PCR probes are available.

In some cases **clonality** of a lymphocytic lesion (neoplastic versus reactive) cannot be proven by FCM, and molecular studies are needed to demonstrate clonal rearrangements of the Ig or T-cell receptor (TCR) genes. Lymphocytes express Ig or TCR genes that are subjected to recombination events during development in the BM and thymus and rearrangements when exposed to foreign or self antigens. A large variety of gene configurations occur due to imperfections in splicing and the insertion of random nucleotides during the recombination events as well as the huge diversity of antigens to which lymphocytes are exposed. When a neoplastic (monoclonal) process arises, there is a shared gene structure that will be observed in a majority of the lymphocytes in the lesion. Although there is a tremendous range of gene configurations that can result from clonal rearrangements, there are **consensus gene sequences** flanking the core rearrangement loci for which PCR probes have been developed. The most common consensus gene sequences exploited for this purpose are the Ig heavy chain and κ light chain genes for B cells and the TCR γ gene for T cells. Alternatively, Southern blot hybridization studies may be used.

Microarrays

References to this and other new **gene profiling** methods are accumulating rapidly in the hematopathology literature, but direct clinical applications have not yet appeared in community hospital practice. However, the potential is tremendous and its **basic principles** deserve discussion. Microarrays consist of up to tens of thousands of oligonucleotide probes placed in separate spots within a small area, usually on a glass slide or silicon chip. These are subjected to the fluorescently labeled genetic material to be tested. Complimentary genes hybridize with the probes on the testing surface, non-hybridized genetic material is washed away, and the slide is placed in a reader. Within the reader, a laser excites each spot. The resulting fluorescence is analyzed by computer. Over- and under-expression of the genes is determined by comparing the disease tissue to normal control tissue. In this way, enormous numbers of genes are analyzed in a very short period of time. Currently, the major problem with microarrays is that the volume of data acquired is outpacing the ability of researchers to meaningfully deduce. Hematologic neoplasms are likely to be one of the early areas in which practical clinical applications are developed for these technologies.

2

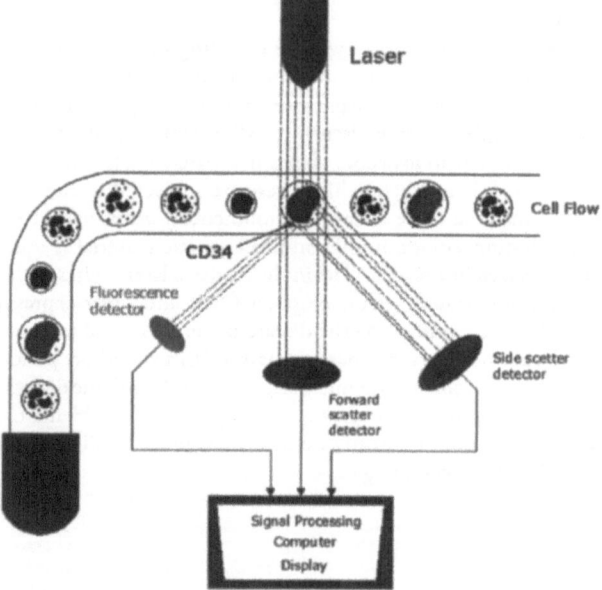

Principle of flow cytometry

Figure 2.1. Simplified diagram of a flow cytometer.

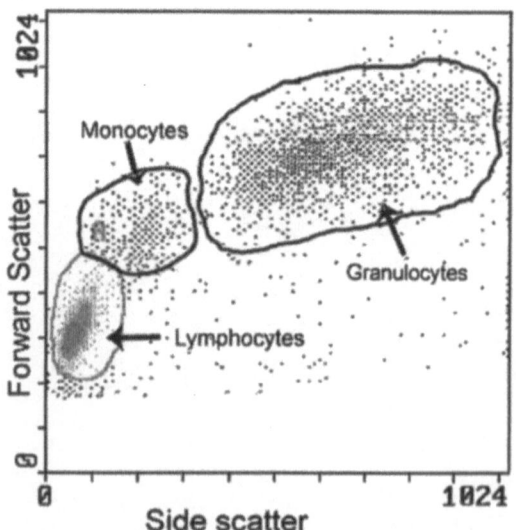

Figure 2.2. Flow cytometry scattergram of peripheral blood showing gates drawn around the different cell populations.

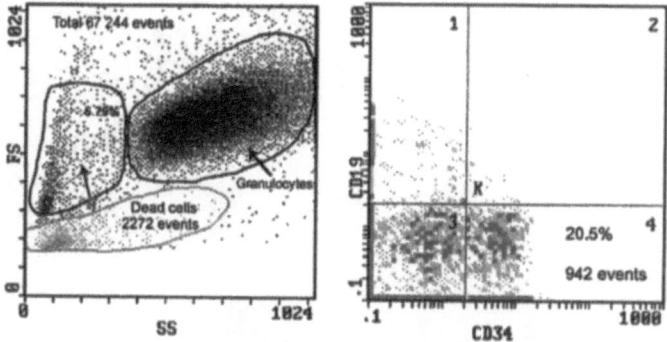

**Calculation of the percentage of
myeloblasts in the bone marrow
by flow cytometry.**

Figure 2.3. Use of flow cytometry to estimate blast percentage by CD34 expression and cluster recognition.

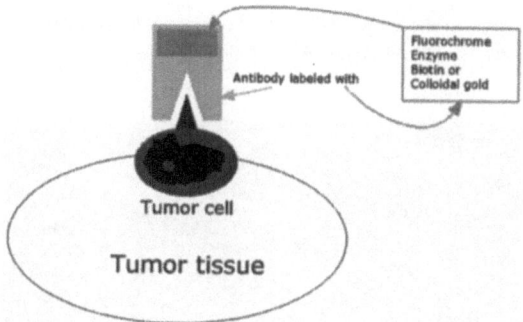

Figure 2.4. Direct method of immunohistochemistry.

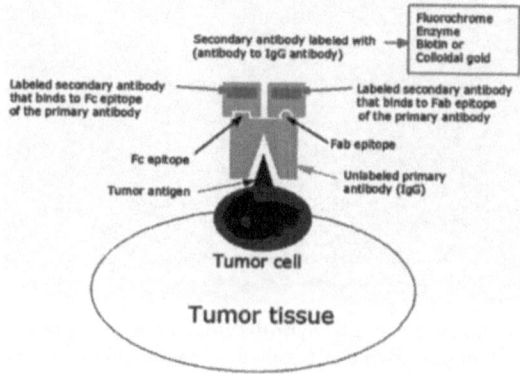

Figure 2.5. Sandwich method of immunohistochemistry.

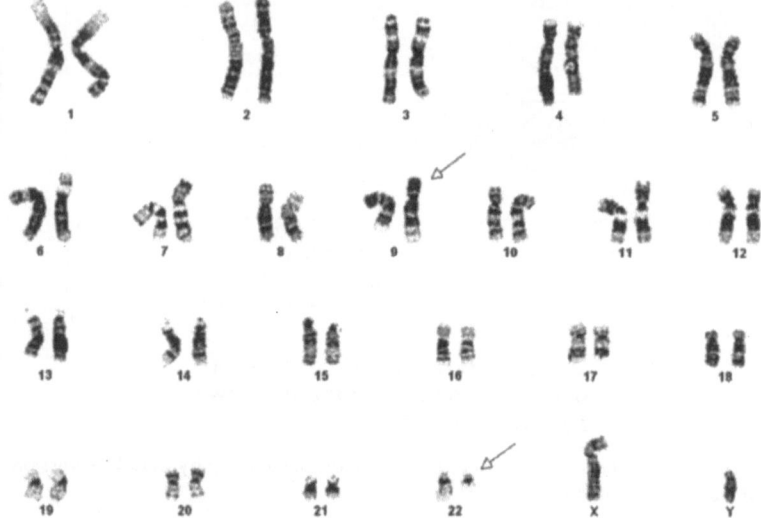

Figure 2.6. Karyogram showing the t(9;22) abnormality (Philadelphia chromosome) (photo courtesy of CSI Laboratories, Alpharetta, GA).

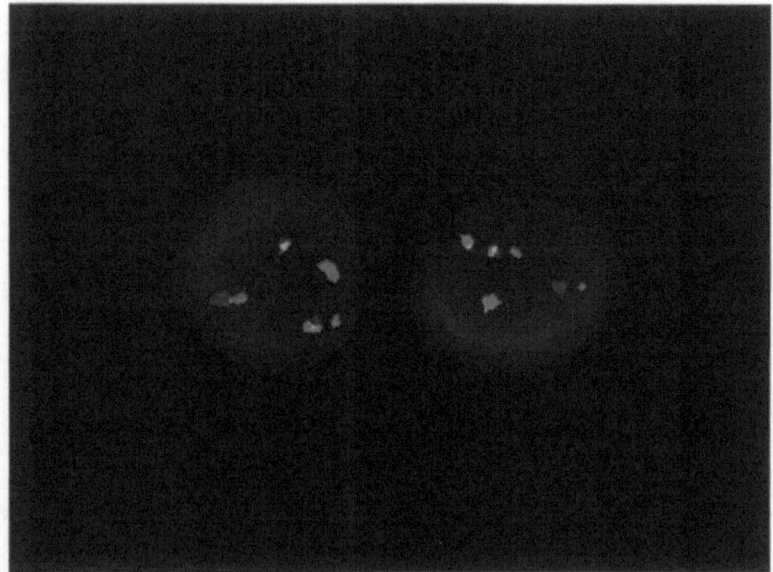

Figure 2.7. Fluorescence in situ hybridization of interphase cells showing BCR(green)/ABL(orange) fusion (Philadelphia chromosome) (photo courtesy of CSI Laboratories, Alpharetta, GA).

Chapter 3

Myelodysplastic Syndromes

Introduction

Myelodysplastic syndromes (MDSs) are clonal disorders of hematopoietic cells that are manifested in the peripheral blood (PB) by cytopenias and clinically by the symptoms resulting from cytopenias or from blood cell dysfunction. The classic triad for diagnosing MDS is:

- PB cytopenia(s)
- Dysplasia in one or more cell line
- Bone marrow (BM) hypercellularity

MDS is most commonly identified in BM specimens performed to evaluate anemia in elderly patients.

Peripheral blood cytopenias are identified by an automated complete blood count (CBC). Anemia is most common, followed by thrombocytopenia and neutropenia. Cytopenias of MDS are usually moderate or mild. Anemia is frequently slightly macrocytic with a normal red blood cell distribution width (RDW).

Clinically:

- Anemia leads to fatigue.
- Thrombocytopenia causes petechiae or bleeding.
- Neutropenia is associated with infections.

MDS is commonly discovered incidentally during routine blood work.

Bone marrow hypercellularity is identified during examination of the biopsy specimen. BM cellularity is approximately 100% at birth and decreases by about 10% per decade. Cellularity can also be estimated from the BM clot sections, but not accurately so from the aspirate.

The real key to the diagnosis of MDS is the ability to recognize dysplasia of one or more of the BM or PB elements. This is based on a small number of simple features.

Erythroid Dysplasia

The erythroid cell line is the most common in which dysplasia is identified. It consists of one or more of three basic morphologic characteristics:

- Nuclear membrane irregularities
- Multinucleation
- Ring sideroblasts

Two accessory morphologic changes are supportive rather than diagnostic for dysplasia:

- Megaloblastoid changes
- Cytoplasmic vacuolization

Bone Marrow: A Practical Manual, by Daniel A. Cherry and Tomislav M. Jelic.
©2011 Landes Bioscience.

Dysplastic nuclear features: Normal erythroid nuclei are round with regular, smooth nuclear membranes. Dysplastic variations include:
- Oval or irregular shape
- Notching or budding of the nuclear membrane
- Small nuclear fragments arranged like satellites around the nucleus

Multinucleation is abnormal. In some instances, there is bridging of nuclear material from one cell to an adjacent cell.

Ring sideroblasts are seen on the iron stain BM aspirate smear. The World Health Organization (WHO) definition requires that 15% or more of the erythroid cells demonstrate the nuclei to be two-thirds or more encircled by siderotic granules. Additional dysplastic features may or may not be present. For this reason, MDS cannot be completely ruled out unless there is adequate iron.

Megaloblastoid change, as opposed to megaloblastic, refers to morphologic changes that resemble those seen with B12 and folate deficiencies in the absence of deficiency or without specific knowledge of the status of these nutrients. The nuclei are enlarged, with stippled chromatin described as looking like pepperoni pizza. Megaloblastoid change easily can be over-diagnosed because of staining artifact.

Cytoplasmic vacuolization of erythroid cells can be difficult to distinguish from artifact and, alone, is not sufficient evidence for dysplasia.

Megakaryocytic Dysplasia

Characterized by:
- Abnormal size
- Abnormal nuclear segmentation
- Cytoplasmic hypogranulation

Like the other cell lines, megakaryocytes demonstrate a range of maturation in the BM, which, in general, is associated with increasing size, nuclear lobation and cytoplasmic granulation. Relatively increased immature forms, a left-shift, may be mistaken for dysplasia.

Cell size varies with maturation, but the majority of megakaryocytes should be two to four times the size of a blast.

Nuclear abnormalities: Dysplastic megakaryocytic nuclei may be hypersegmented and/or bizarre in appearance, or they may be hyposegmented. Dysplastic hyposegmentation should be differentiated from that seen in immature megakaryocytes, which may be increased in number to compensate for the peripheral destruction of platelets (idiopathic thrombocytopenic purpura, hypersplenism). Complete separation of nuclear segments is abnormal and is seen by focusing up and down. Sectioning can artifactually cause this appearance on the BM biopsy.

Cytoplasmic hypogranulation may accompany abnormalities of size and nuclear morphology. Asynchronous granulation—that is, a mature-appearing megakaryocyte lacking cytoplasmic granulation, is usually dysplastic. Sometimes large agranular platelets are seen in the PB. Dysplasia is infrequently diagnosed based on hypogranulation alone.

Granulocytic Dysplasia

Most frequently seen in the PB, but may also be appreciated in the BM aspirate smear. Takes the form of:

- Nuclear hyposegmentation or hypersegmentation
- Abnormalities of cytoplasmic granulation

Hyposegmented granulocytes, either bilobed or unilobed, are commonly referred to as **pseudo-Pelger-Huët cells**. **True Pelger-Huët anomaly** is usually not identified for the first time in the same clinical setting as MDS: an older or elderly patient being worked up for persistent cytopenia(s). True Pelger-Huët anomaly is more uniform in appearance from cell to cell and is seen in virtually every neutrophil. Sometimes ring-form nuclei are seen in dysplastic granulocytes.

Hypogranulation and Auer rods: Granulocytic hyposegmentation is accompanied by hypogranulation frequently and, less commonly, by granular polarization. Auer rods, a form of abnormal granulation, are never seen in normal granulocytes. In some settings, their presence upgrades the classification of the MDS (see classification of MDS below).

Hypersegmentation may represent dysplasia but is difficult to differentiate from the megaloblastic changes observed with deficiencies of B12 or folate.

BM Changes Supportive of MDS

Two BM findings unrelated to dysplasia in individual cell lines support the diagnosis of MDS but are insufficient in the absence of dysplasia.

Abnormal localization of immature myeloid precursor cells (ALIP) is defined as the presence of at least three groups of at least three to more than five immature myeloid cells (myeloblasts and promyelocytes) residing in the spaces between the BM trabeculae, rather than their normal location hugging the trabeculae. Several such groups must be identified because they may falsely appear to be intertrabecular when a trabeculum actually lies just beneath the plane of sectioning. These cells must also be shown to be myeloid and not immature erythroid or megakaryocytic cells. This can be difficult based on morphology alone and is most easily demonstrated by positive immunohistochemical (IHC) staining with myeloperoxidase. When present, ALIP is strong evidence for MDS.

Reticulin fibrosis: Mild (1+ to 2+) reticulin fibrosis demonstrated by a special stain may be seen in the BM biopsy specimen in MDS. Grading of reticulin fibrosis is elucidated in the discussion of myeloproliferative neoplasms.

Hypoplastic MDS

Myelodysplastic syndromes almost always demonstrate BM hypercellularity, but there are rare exceptions known as hypocellular or hypoplastic MDS. The diagnosis of hypoplastic MDS is heavily dependent on identifying ALIP in the BM biopsy sections.

Hypoplastic MDS versus aplastic anemia (AA): There is considerable overlap of the morphologic and clinical features of hypoplastic MDS and AA and some controversy as to whether they are separate entities. For this reason, and because of the general difficulty of making this diagnosis, great care should be taken to identify dysplasia in such cases. An IHC stain for CD20 should always be done to rule out hairy cell leukemia (HCL).

Table 3.1. Features of WHO classes of MDS

MDS	Blasts in PB	Blasts in BM	Cellular Features
RCUD (RA, RN, RT)	<1%	<5%	dysplasia in one cell line only; may have bicytopenia; if pancytopenic, then classify as MDS, U
RARS	<1%	<5%	≥15% ring sideroblasts with or without erythroid dysplasia but no significant granulocytic or megakaryocytic dysplasia
RCMD	<1%	<5%	dysplasia in ≥10% of cells in two or more cell lines; If 1% blasts persistently present in PB, then classify as MDS, U
RCMD-RS	<1%	<5%	Same as RCMD plus ≥15% ring sideroblasts
RAEB-1	≤1% with BM blasts 5-9% Or 2-4% with BM blasts <5%	5-9% with PB blasts ≤1% Or <5% with PB blasts 2-4%	Unilineage or multilineage dysplasia
RAEB-2	5-19%	10-19%	Unilineage or multilineage dysplasia

RCUD: refractory cytopenia with unilineage dysplasia; RA: refractory anemia; RN: refractory neutropenia; RT: refractory thrombocytopenia; MDS, U: myelodysplastic syndrome, unclassified; RARS: refractory anemia with ring sideroblasts; RCMD: refractory cytopenia with multilineage dysplasia; RCMD-RS: RCMD with ring sideroblasts; RAEB: refractory anemia with excess blasts. Adapted with permission from: Swerdlow SH, Campo E, Harris NL et al. World Health Organization Classification of Tumours of Haematopoietic and Lymphoid Tissues. Lyon: IARC, 2008:89, Table 5.01.

Classification of MDS

The WHO classification scheme is based on which and how many cell lines demonstrate dysplasia, the number of blasts and a few special features such as the presence of ring sideroblasts and Auer rods. Table 3.1 summarizes the WHO classification of MDS.

There are two additional classes of MDS, as well as a provisional entity not depicted in the above chart: MDS, unclassified (MDS, U), 5q- syndrome, and refractory anemia with ring sideroblasts and thrombocytosis (RARS-T).

MDS, U: This diagnosis is used in cases that do not fit into one of the other categories. There are three specific situations in which this diagnosis is used:

- RCUD or RCMD, in which 1% blasts are found in the PB on two separate occasions
- RCUD accompanied by pancytopenia

Table 3.2. MDS survival statistics and risks for the development of acute leukemia

MDS	Median Survival	Progression to Acute Leukemia
RA	66 months	6%
RARS	72 months	1-2%
RCMD	33 months	11%
RAEB-1	18 months	25%
RAEB-2	10 months	33%
5q- syndrome	Long	Rare
MDS unclassified	Unknown	Unknown

3

- Persistent cytopenias without morphologic evidence of dysplasia, or MDS with typical morphologic features in less than 10% of cells but with a chromosomal abnormality that is associated with MDS (see section on genetics and Table 3.4).

5q- syndrome: MDS associated with the 5q- chromosomal abnormality, which can be identified by standard cytogenetic studies or by fluorescence in situ hybridization (FISH), demonstrates unique features. There is usually PB thrombocytosis and BM megakaryocytic hyperplasia with hypolobated nuclei. It is defined by the del(5q) chromosomal abnormality. These cases have an excellent prognosis and usually contain <5% blasts in the BM and <1% blasts in the PB. Cases with ≥5% blasts in the BM or ≥1% blasts in the PB usually possess additional chromosomal abnormalities and are associated with a poorer prognosis. Auer rods are not present. This neoplasm is highly sensitive to the drug lenalidomide.

RARS-T is a provisional entity in which RARS is accompanied by a platelet count >450 x 10^9/L with megakaryocyte morphology similar to that seen with essential thrombocytosis, large and hypersegmented (see Myeloproliferative Neoplasms).

Auer rods are caused by the alignment of cytoplasmic granules. Their significance in MDS is not fully known. The WHO classification recommends that cases of RAEB-1 with Auer rods be classified as RAEB-2, but it does not explain their role in the classification of other MDSs.

Prognosis

Major risk factors: The prognosis for MDS patients is expressed as either median survival or as the rate of progression to acute leukemia. The two major predictors are the number of blasts and cytogenetic abnormalities. Their relationships to prognosis are depicted in Tables 3.2 and 3.3.

Additional risk factors related to poor prognosis include pancytopenia and age over 60.

Table 3.3. MDS prognostic categories based on cytogenetic abnormalities

Good Risk	Poor Risk	Intermediate Risk
Normal genetics or isolated del(5q) or Y-	Complex abnormalities (three or more recurring abnormalities), or abnormalities of chromosome 7	All other abnormalities

International Prognostic Scoring System (IPSS): Some institutions use the IPSS, which is based on blast count, cytogenetic findings and cytopenias. An IPSS is customarily assigned by the clinician rather than by the pathologist. Details of the IPSS can be found in the WHO bluebook.

Special Studies in MDS

The results of all special studies should be incorporated into the BM report, regardless of whether they have separate reports. This is commonly accomplished by addendums, since the case is usually signed out before the results are received from the reference laboratory. The addendum need not contain all of the detail given in the reference laboratory report, but it should include sufficient information so that anyone reading the BM report does not need to separately obtain the reference laboratory report for clinically relevant information.

Cytogenetics (chromosomal analysis) is currently the most important special study for the diagnosis of MDS. The identification of a chromosomal abnormality strongly supports the diagnosis of MDS and has important prognostic implications, as described above. Chromosomal abnormalities commonly associated with MDS are given in Table 3.4. When they occur as single abnormalities such as +8, del(20q) and –Y, they are not considered to be presumptive of MDS unless other findings are present (e.g. cytopenias, dysplasia).

Table 3.4. Chromosomal abnormalities associated with MDS, in decreasing order of frequency

---

Abnormality
+8
-7 or del(7q)*
-5 or del(5q)*
del(20q)
-Y
i(17q) or t(17p)
-13 or del(13q)
del(11q)
del(12p) or t(12p)
del(9q)
idic(x)(q13)
t(11;16)(q23;p13.3)*
t(3;21)(q26.2;q22.1)*
t(1;3)(p36.3;q21.2)
t(2;11)(p21;q23)
inv(3)(q21;q26.2)
t(6;9)(p23;q34)

---

* - have a specific association with treatment-related MDS (T-MDS). See *Acute Myeloid Leukemia* chapter. Adapted with permission from: Swerdlow SH, Campo E, Harris NL et al. World Health Organization Classification of Tumours of Haematopoietic and Lymphoid Tissues. Lyon: IARC, 2008:93, Table 5.04.

FISH probes are available for the common chromosomal abnormalities associated with MDS. Many reference laboratories now offer MDS FISH panels that include multiple probes for the more important of these. Advantages of FISH over standard cytogenetics are that it can be done on archived paraffin-embedded clot sections, the results are available more quickly, and sensitivity is superior. Cytogenetics is sufficiently sensitive to detect these abnormalities in most instances, and FISH is rarely indicated.

Flow cytometry (FCM) can be helpful in some cases of MDS. Most importantly, FCM offers a cross-check of the blast count by morphology, but they do not correlate perfectly. Demonstration of a left-shift (i.e., increased blasts) is the most specific FCM abnormality associated with MDS. Additionally, FCM in MDS can show abnormalities of the granulocytic maturation sequence, aberrant antigen expression and increased hematogones, none of which are entirely specific. The presence of a single abnormal feature is felt to be insignificant. As technical improvements are made, along with advances in our understanding, FCM has the potential to become very important in the evaluation of MDS. At this time, however, FCM plays a secondary role.

IHC stains of the BM biopsy sections can be helpful for differentiating immature myeloid-form erythroid and megakaryocytic cells. A simple battery of stains including myeloperoxidase, glycophorin A and Factor VIII (or CD61) is an easy way of identifying these cell lines, respectively.

Initial Diagnosis of MDS

A diagnosis of MDS can be made on the initial BM sample in cases with increased blasts (RAEB), Auer rods, or when a chromosomal abnormality is present. However, it is a frustrating fact that many cases lack definitive evidence for MDS but are suspicious. Common problems include:

- Mild morphologic changes, insufficient for diagnosing dysplasia
- Dysplasia in less than 10% of the cells
- The absence of genetic data because of the long turnaround time for chromosomal studies
- Suboptimal specimens due to air drying artifact or hypocellularity

The BM report should explain such factors and express an appropriate element of ambiguity. Dysplasia can be due to etiologies other than clonal myeloid disorders (Table 3.5). Caution against over-diagnosis should always be practiced, since MDSs are chronic diseases in which moderate delays in treatment are rarely clinically significant.

Table 3.5. Non-MDS causes of dysplasia

Causes of Dysplasia Other Than MDS
Chemotherapy
Radiation therapy
Heavy metal exposure (especially arsenic; sometimes present in well water)
Non-hematologic malignancies involving the BM
Viral infections (HIV, Parvovirus and others)
Nutritional deficiencies (B12, folic acid)
Congenital diseases (congenital dyserythropoietic syndromes)

Waiting period for the definitive diagnosis of RA: The WHO bluebook states, "If a clonal abnormality is not present, there should be an observation period of six months, before a diagnosis of RA is established." The major disadvantages of over-diagnosing RA are depriving the patient of treatment for another cause of anemia and the undue anxiety caused by the diagnosis of a neoplasm. The initial treatments for RA are usually nonaggressive, such as erythropoietin administration, and pose little risk. Nevertheless, under-diagnosis is preferable.

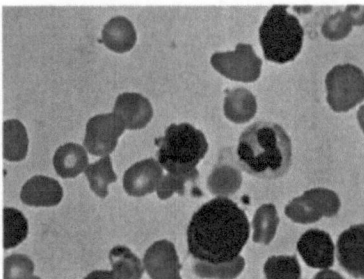

Figure 3.1. Dysplastic erythroid cell with irregular nuclear membrane (BM aspirate, 500x).

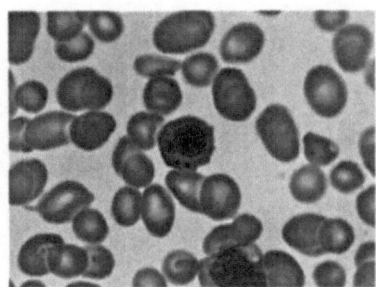

Figure 3.2. Dysplastic erythroid cell with satellite nuclear fragments (BM aspirate, 500x).

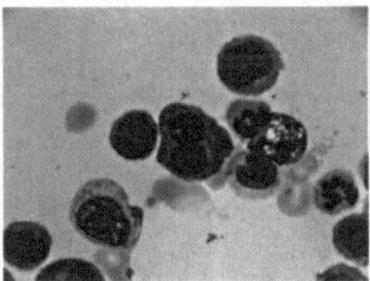

Figure 3.3. Biclucleated erythroid cell (BM aspirate 500x).

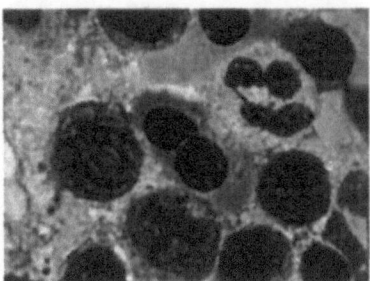

Figure 3.4. Dysplastic erythroid cell with nuclear bridge (BM aspirate, 500x).

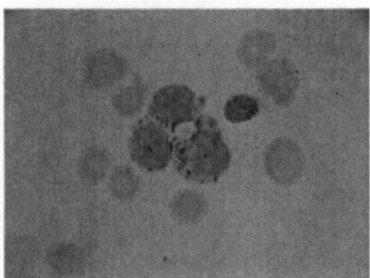

Figure 3.5. Ring sideroblasts (iron stained BM aspirate 500x).

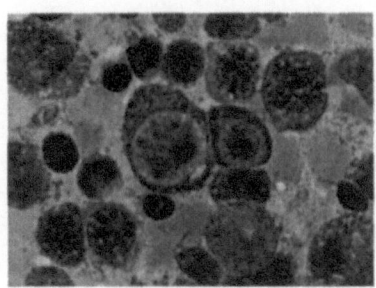

Figure 3.6. Megaloblastoid change of erythroid cells (BM aspirate 500x).

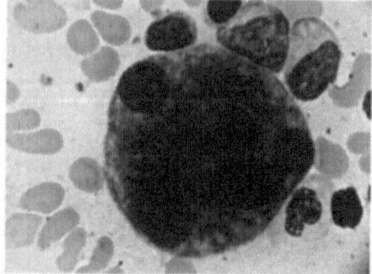

Figure 3.7. Dysplastic megakaryocyte with complete separation of nuclear lobes (BM aspirate, 500x).

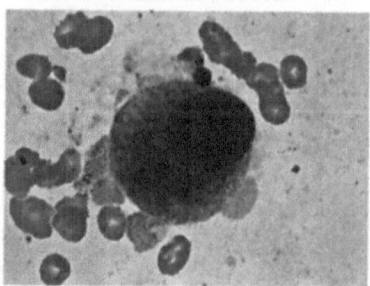

Figure 3.8. Uninuclear megakaryocyte (BM aspirate, 500x).

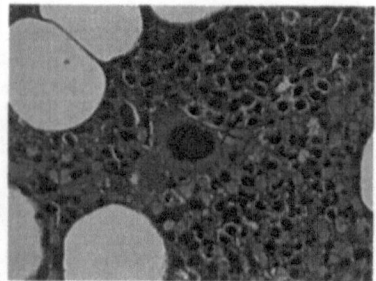

Figure 3.9. Hypolobated megakaryocyte (BM biopsy, 400x).

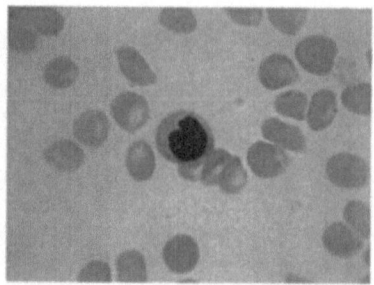

Figure 3.10. Hyposegmented and hypogranular neutrophil (PB, 500x).

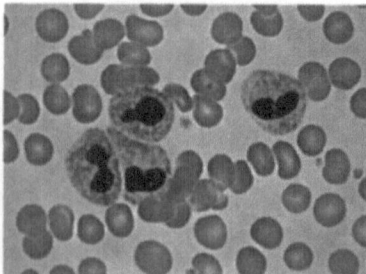

Figure 3.11. Pseudo-Pelger-Huët cells (PB, 500x).

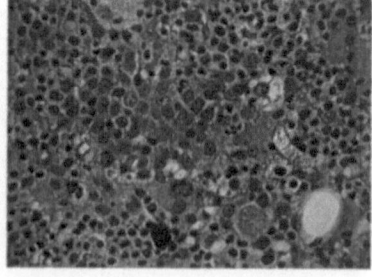

Figure 3.12. Abnormal localization of immature myeloid precursor cells (ALIP) (BM biopsy, 200x).

CHAPTER 4

Myeloproliferative Neoplasms

Introduction

Chronic myeloproliferative neoplasms (MPNs) are clonal disorders of the hematopoietic stem cells that result in elevation of one of the myeloid cell lines: granulocytic, erythroid or megakaryocytic. In the early phases of these diseases, the affected cell lines maintain their function and their ability to mature. In other words, hematopoiesis remains "effective." In later phases, clonal evolution results in "ineffective" hematopoiesis; that is, loss of the ability of the affected cell line(s) to function and mature, leading to bone marrow (BM) failure that is sometimes exacerbated by reactive BM fibrosis. Some cases progress to the final stage of clonal evolution-acute leukemia, usually myeloid but sometimes lymphoid. Formerly known as myeloproliferative disorders, the revised World Health Organization (WHO) classification system has renamed them because they are now uniformly accepted to be neoplastic.

Presentation and findings: Because hematopoiesis remains "effective" during the early periods of the disease, it is typical for patients to be asymptomatic at presentation and to come to the attention of their physician because of abnormal routine laboratory work or physical examination. When symptoms are present, they are usually due to elevated cell counts or splenomegaly:

- Hypereosinophilic syndrome
- Thrombosis secondary to thrombocytosis
- Plethora or pruritis related to erythrocytosis
- Feeling of fullness or "tugging" sensation in the abdomen because of splenomegaly

In addition, BM hyperproliferation can cause fever, fatigue and weight loss, sometimes referred to as myeloproliferative symptoms.

In cases of primary myelofibrosis (PMF), the patient may present with cytopenias and/or a leukoerythroblastic blood smear with teardrop cells (dacrocytes).

Age and incidence: MPNs occur most frequently in the fifth to seventh decades of life, with an incidence of approximately six to nine new cases per 100,000 in population each year. Most studies show a slight male predominance.

Role of the pathologist: Elevated PB cell counts are the most commonly stated reason for BM sampling. The principal tasks for the pathologist are as follows:

- Differentiate MPN from a reactive etiology
- Classify the MPN

It is unusual for a clinical hematologist to do a BM unless he feels confident that reactive causes have been clinically ruled out to a reasonable extent or the cell counts are very high. Despite this presumed diligence, it is prudent for the pathologist to review all available information relevant to ruling out a reactive

---

Bone Marrow: A Practical Manual, by Daniel A. Cherry and Tomislav M. Jelic.
©2011 Landes Bioscience.

etiology. If additional laboratory work would be helpful, such as an erythropoietin (EPO) level in a patient with erythrocytosis, then the pathologist should order it.

BCR/ABL and JAK2 testing: Chronic myelogenous leukemia (CML) and polycythemia vera (PV) are diagnosed primarily by the identification of the BCR/ABL and JAK2 chromosomal abnormalities, respectively. CML is defined by the presence of the Philadelphia chromosome (BCR/ABL), and the definitions of all other MPNs require its absence. The JAK2 abnormality is now recognized to be present in virtually all cases of PV but may also be present in the other non-CML MPNs and, when positive, rules out a reactive etiology. Therefore, the first step of the BM work-up in every case of suspected MPN is to initiate testing for these genes.

Testing for the BCR/ABL and JAK2 abnormalities is done by fluorescence in situ hybridization (FISH) and polymerase chain reaction (PCR), respectively. Chromosomal analysis (cytogenetics) fails to identify BCR/ABL in 10% of cases. Routine cytogenetics (CG) should also be done to evaluate for unsuspected chromosomal abnormalities such as those that define the accelerated phase of CML.

Reticulin stain: In addition to genetic testing, a special stain for reticulin should be done on the BM biopsy in all cases because a significant element of fibrosis may be morphologically inapparent on routine staining. Reticulin fibrosis can be graded as follows:

- Grade 1: <25% of areas examined occupied by reticulin fibers
- Grade 2: 25-50% of areas examined occupied by reticulin fibers
- Grade 3: >50-75% of areas examined occupied by reticulin fibers
- Grade 4: >75% of areas examined occupied by reticulin fibers

OR by the WHO semiquantitative method:*

- MF-0: Scattered linear reticulin with no intersections (crossovers), corresponding to normal BM
- MF-1: Loose network of reticulin with many intersections, especially in perivascular areas
- MF-2: Diffuse and dense increase in reticulin with extensive intersections, occasionally with focal bundles of collagen and/or osteosclerosis
- MF-3: Diffuse and dense increase in reticulin with extensive intersections and coarse bundles of collagen, often associated with osteosclerosis

Myeloproliferative neoplasm classification is based on BCR/ABL and JAK2 testing in cases of CML and PV, and by the morphological features for the other MPNs. If testing for BCR/ABL is positive, then the diagnosis is CML, excluding instances of acute leukemia, differentiated by a blast count of ≥20 percent. When BCR/ABL is negative, then the clinical, morphologic and laboratory features of the disease must be evaluated and matched with the WHO criteria.

WHO criteria are available for most of the major diagnostic categories of MPNs. Difficulty arises when an MPN is felt to be present but it cannot be satisfactorily placed into one of the defined categories. In some such cases the category of MPN, unclassified (MPN, U), may be used (discussed in detail below). If features of myelodysplasia are also present, then a myelodysplastic/myeloproliferative neoplasm (MDS/MPN) should be contemplated.

*WHO semiquantitative method was adapted with permission from: Swerdlow SH, Campo E, Harris NL et al. World Health Organization Classification of Tumours of Haematopoietic and Lymphoid Tissues. Lyon: IARC, 2008:46, Table 2.05.

Table 4.1. Oversimplified scheme as starting place for MPN classification

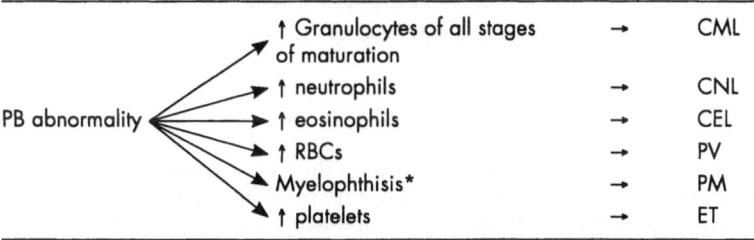

PB abnormality	↑ Granulocytes of all stages of maturation	→	CML
	↑ neutrophils	→	CNL
	↑ eosinophils	→	CEL
	↑ RBCs	→	PV
	Myelophthisis*	→	PM
	↑ platelets	→	ET

CEL: Chronic eosinophilic leukemia; PM: Primary myelofibrosis; CML: Chronic myeloid leukemia; CNL: Chronic neutrophilic leukemia; ET: Essential thrombocythemia; PV: Polycythemia vera; *Myelopthisic blood smear: leukoerythroblastosis and teardrop cells.

It is a gross oversimplification to state that classification is based on which cell line is elevated in the PB (Table 4.1) but this is a good place to start. After all, an abnormal CBC most likely initiated the original referral of the patient to the clinical hematologist.

Philadelphia Chromosome (BCR/ABL)

BCR/ABL results from the t(9;22)(q34;q11) chromosomal abnormality and is present in 90-95% of cases of CML at the time of diagnosis. The remaining cases have variants of this abnormality, some of which are cryptic, meaning they cannot be detected by routine chromosomal analysis (CG). Variations in the breakpoint on the BCR (breakpoint cluster region) gene result in abnormal fusion proteins of different sizes: M-BCR, μ-BCR and m-BCR.

M-BCR: The most common in CML, is the "major breakpoint cluster region" and results in the abnormal fusion protein p210.

μ-BCR: Rarely, the breakpoint occurs in the mu region resulting in the p230 fusion protein, which is sometimes associated with marked neutrophil maturation (CML-N).

m-BCR: The most common form of the Philadelphia chromosome seen with acute lyphoblastic leukemia (ALL) is associated with the p190 fusion protein resulting from the minor breakpoint cluster region present in small amounts in approximately 90% of cases of CML and is very rarely associated with cases of CML that demonstrate increased numbers of monocytes.

BCR/ABL as a screening test: The presence of BRR/ABL in normal individuals may indicate an increased risk of developing CML in the future, but this risk is not 100% and it is not known that any intervention can prevent the development of CML in these individuals. Testing for BCR/ABL should not be done unless there is specific clinical or laboratory evidence of an MPN or ALL.

JAK2 Gene

A homozygous mutation resulting in the substitution of phenylalanine for valine at position 617 of the Janus kinase 2 (*JAK2*) gene, referred to as the V617F mutation or sometimes the *VF* gene, has been identified in large numbers of

Table 4.2. Frequency of JAK2 in non-CML MPNs

PV	81-97% (remaining are positive for the exon 12 variant)
PM	43-50%
ET	41-57%

patients with MPNs (Table 4.2). Testing for the V617F mutation of *JAK2* is recommended in all cases, because its presence rules out reactive etiologies for erythrocytosis and thrombocytosis and non-MPN causes of BM fibrosis. In cases of erythrocytosis, identification of *JAK2* is one of the major criteria for the diagnosis of PV.

Chronic Myelogenous Leukemia, BCR-ABL Positive

CML is a clonal disease of the hematopoietic stem cells that is manifested primarily as neutrophilic granulocytosis. CML is defined by the presence of the Philadelphia chromosome, which may be found in all of the myeloid cell lines and some lymphoid and epithelial cells. Characteristic disease phases include an initial chronic phase (CML-CP), which without treatment evolves to the more aggressive accelerated (CML-AP) and blast (CML-BP) phases.

Worldwide, CML occurs in about 1 to 2 per 100,000 persons with a median age of 50 years and a slight male predominance.

Clinical: Approximately 20-40% of patients are asymptomatic at the time of presentation and are discovered on routine blood counts. Other common presenting features are lethargy, weight loss, bleeding and splenomegaly, which may cause the patient to complain of a feeling of fullness or a "tugging" sensation in the left upper abdomen. Less common presentations include night sweats, lymphadenopathy, bone pain, infection, hyperviscosity and thyrotoxicosis. Rarely, patients present in accelerated or blast phase.

Peripheral blood findings in CML-CP include the following:
- Leukocytosis ranging from 12 to 1000 x 10^9/L with a mean of about 100 x 10^9/L; comprised almost entirely of neutrophils and neutrophil precursor cells of all stages of maturation with few blasts (<10%) and a predominance of cells in the mid-maturation stages, sometimes called a "myelocytic bulge."
- Absolute basophilia in 100% of cases
- Absolute eosinophilia in 90%
- Frequently, mild normochromic/normocytic anemia
- Frequently, elevated platelets that may be so elevated (1,000 x 109/L or more) as to cause confusion with ET
- Thrombocytopenia is rare.
- No significant dysplasia

Bone marrow findings in CML-CP (Fig. 4.2, 4.3)
- Hypercellularity approaching 100% with a composition similar to that of the PB
- Predominately neutrophilic cells with few blasts and a myelocytic bulge as well as increased eosinophilic and basophilic cells; paratrabecular cuff of myeloid cells is 5 to 10 cells in thickness, as opposed to the normal 2 to 3 cells thick.

Table 4.3. WHO criteria for CML-AP

The WHO states that CML-AP is present if one or more of the following findings are present:
  1. Persisting or increasing WBC (>10 x 10⁹/L) and increasing spleen size unresponsive to therapy
  2. Persistent thrombocytosis (>1,000 x 10⁹/L) unresponsive to therapy
  3. Persistent thrombocytopenia (<100 x 10⁹/L) unrelated to therapy
  4. Cytogenetic evidence of clonal evolution*

*See Cytogenetic testing in text. Adapted with permission from: Jaffe ES, Harris NL, Stein H, Vardiman JW. WHO Classification of Tumours, Pathology and Genetics of Tumours of Haematopoietic and Lymphoid Tissues. Lyon: IARC, 2001:21, Table 1.03.

- Megakaryocytes increased in number and small ("dwarf megakaryocytes"), hypolobated, and sometimes clustered. This morphology is helpful in differentiating CML from ET in which the megakaryocytes are large or giant with deeply lobated nuclei (Fig. 4.4).
- Reticulin fibrosis
- Abnormal macrophages frequently present, including pseudo-Gaucher cells and "sea-blue" histiocytes
- No significant dysplasia

Involvement of other organs: Leukemic involvement of the liver and spleen may also be seen, but these tissues are rarely sampled for evaluation.

Laboratory findings: Other important laboratory findings include the following:
- Increased B12 levels (may be 10 to 20 times normal)
- Elevated uric acid

Leukocyte alkaline phosphatase score: An older test used for the differentiation of leukemoid reactions, a catch-all phrase for non-neoplastic elevation of granulocytes, from CML is the leukocyte (or neutrophil) alkaline phosphatase (LAP or NAP) score. In this test the PB is cytochemically stained for LAP, then 200 neutrophils are scored for positivity on a scale of 0 to 4+ and the scores are totaled. A decreased score, usually <13, is associated with CML, rather than a leukemoid reaction. Although the LAP score is rapid and useful, it is seldom used anymore. Technicians proficient at performing this test are becoming scarce.

Chronic myeloid leukemia, accelerated phase signifies an increase in aggressiveness of disease. The WHO has designated specific criteria for CML-AP, given in Table 4.3. Other evidence of accelerated phase that have not been well tested in clinical trials includes marked granulocytic dysplasia or prominent proliferation of small dysplastic megakaryocytes in large clusters or sheets associated with marked reticulin or collagen fibrosis. The presence of lymphoblasts in the PB or BM is also a cause for concern.

Chronic myeloid leukemia, blast phase: The WHO definition for the blast phase of CML is given in Table 4.4. This is usually myeloid (50-60% of cases) but may be lymphoid (16-30% of cases), erythroid, megakaryocytic, biphenotypic or bilineal. In approximately 5-10% of cases, an extramedullary blast proliferation may occur in such sites as lymph nodes (LNs), soft tissues or central nervous system

Table 4.4. WHO criteria for CML-BP

The WHO states that CML-BP may be diagnosed if one or more of the following findings is identified:
   1. Blasts 20% or more of PB WBCs and/or BM nucleated cells
   2. Extramedullary blast proliferations

Adapted with permission from: Jaffe ES, Harris NL, Stein H, Vardiman JW. WHO Classification of Tumours, Pathology and Genetics of Tumours of Haematopoietic and Lymphoid Tissues. Lyon: IARC, 2001:23, Table 1.04.

(CNS) and may be myeloid or lymphoid. Extramedullary blast proliferations may occur before, simultaneous to or after BM blast proliferation.

Cytogenetic testing is done to evaluate for clonal evolution in the form of the appearance of additional chromosomal abnormalities such as these:
- Additional Philadelphia chromosome
- (+)8
- i(17q)
- (+)19

If acquired after the time of original diagnosis, these are associated with disease progression. The significance of these findings at the time of original diagnosis is uncertain but presumably signifies increased disease aggressiveness.

Use of special testing to follow treatment: The presence or absence of BCR/ABL is also used in various schemes to follow disease in patients. Some of the most commonly used protocols are those recommended by the drug manufacturers.

Differential diagnosis: The most difficult differential diagnosis of CML is that of a leukemoid reaction. The definitive abnormality that differentiates CML, not only from leukemoid reactions but also from all other diseases with features similar to CML, is the presence of BCR/ABL. Unfortunately, this information may not be immediately available. Other useful features for the differentiation of CML from a leukemoid reaction are given in Table 4.5. Table 4.6 gives a list of other differentials with brief comments about distinguishing characteristics helpful in the early evaluation of these cases.

Prognosis: Therapy with imatinib achieves a complete cytogenetic remission rate of 70-90% with a five-year overall survival of 80-95%.

Chronic Neutrophilic Leukemia

Chronic neutrophilic leukemia (CNL) is an extremely rare disease defined by the following:
- Sustained PB neutrophilia
- Hypercellular BM due to increased neutrophils
- Hepatosplenomegaly
- No BCR/ABL
- No reactive etiology for the neutrophilia

About 150 cases had been reported in the literature. Most patients are elderly with no definite sex predominance. The WHO criteria for the diagnosis of CNL are depicted in Table 4.7.

Table 4.5. Useful features for differentiating leukemoid reaction from CML

	Leukemoid Reaction	CML
PB	WBC usually <100 x 10⁹/L	WBC commonly >100 x 10⁹/L
	Mostly neutrophils with moderate numbers of bands and few meta-myelocytes, myelocytes and blasts	Granulocytes of all levels of maturation with myelocytic bulge
	Toxic granules, cytoplasmic vacuoles, Döhle bodies frequently present	Reactive features in neutrophils absent
	Usually lack significant basophilia or eosinophilia	Absolute basophilia in 100% and eosinophilia in 90%
	NRBCs not present	NRBCs present
BM	Cellularity not markedly increased	Cellularity near 100%
	Increased normal-appearing megakaryocytes	Increased small hypolobated megakaryocytes
	No reticulin fibrosis	Slight-to-marked reticulin fibrosis
	Lack abnormal macrophages	Pseudo-Gaucher cells and sea-blue histiocytes
Clinical	Obvious underlying reactive etiology	No apparent reactive etiology
	Splenomegaly rare	Splenomegaly common

Table 4.6. Differential diagnosis of CML

Disease	Features that Differ from CML
CMML	Monocytes usually >10%; basophils <2%; dysplasia present
CNL	Predominance of mature neutrophils; ↑ LAP score; extremely rare disease
CEL	Predominance of eosinophils; clinical symptoms of hypereosinophilia
PV	Erythrocytosis; clinical features of erythrocytosis; ↓ EPO
PM	Prominent reticulin fibrosis of BM; myelopthisic PB smear
ET	Platelet count usually >1000 x 10⁹/L; large megakaryocytes with hyperlobation
ALL	Lymphoblasts ≥20%; de novo ALL lacks multilineage involvement by disease, as opposed to lymphoid blast phase of CML
AML	Myeloblasts ≥20%; most cases of BCR/ABL⁺ AML are probably CML presenting in blast phase

Table 4.7. WHO criteria for CNL

1. Peripheral blood leukocytosis (WBC ≥25 x 10⁹/L)
   - Segmented neutrophils and bands >80% of WBCs
   - Immature granulocytes (left of band stage) <10% of WBCs
   - Myeloblasts <1% of WBCs
2. Hypercellular BM biopsy
   - Neutrophils increased in percentage and number
   - Myeloblasts <5% of nucleated BM cells
   - Neutrophilic maturation pattern normal
   - Megakaryocytes normal or left shifted
3. Hepatosplenomegaly
4. No identifiable cause for physiologic neutrophilia or, if present, demonstration of myeloid clonality by CG or molecular studies
   - No infectious of inflammatory process
   - No underlying tumor
5. No BCR/ABL
6. No rearrangement of PDGFRA, PDGFRB or FGFR1
7. No evidence of PV, PM or ET
8. No evidence of MDS or MDS/MPN
   - No granulocytic dysplasia
   - No myelodysplastic changes in other myeloid lineages
   - Monocytes <1 x 10⁹/L

Adapted with permission from: Swerdlow SH, Campo E, Harris NL et al. World Health Organization Classification of Tumours of Haematopoietic and Lymphoid Tissues. Lyon: IARC, 2008:39, Table 2.01.

Clinical findings of CNL include the following:
- Hepatosplenomegaly
- Bleeding from mucocutaneous surfaces
- Gout
- Pruritis

The hepatosplenomegaly is due to leukemic infiltration by neutrophils. Any soft tissue may be, likewise, involved by leukemic neutrophil infiltrates. The morphologic features of this disease are encompassed in the WHO diagnostic criteria.

Genetics: Most cases lack cytogenetic abnormalities, but some cases possess +8, +9, del(20q), del(11q) and del(12p). Rare JAK2 positive cases have been reported.

Chronic myeloid leukemia with PB neutrophilia: Cases with similar morphologic features that demonstrate BCR/ABL are classified as a CML variant (CML-N).

Prognosis is variable with survival ranging from 6 months to more than 20 years. Most cases are slowly progressive. Transformation to AML may occur.

Polycythemia Vera

Polycythemia vera (PV) is a clonal disorder of the hematopoietic stem cell that results in increased RBC production. The incidence is approximately 8 to 10 new cases per 1 million in population per year. The average patient is in his 60s, although a wide range of ages may be affected, and there is a slight male predominance in most studies.

Table 4.8. Symptoms and physical findings in PV

Symptoms
Headaches
Dizziness
Scotomata
Myocardial infarction
Stroke
Angiogenic pruritis
Parasthesias
Erythromelalgia
Deep vein thrombosis
Mesenteric, portal, or splenic thrombosis
Gout
Hemorrhage (especially gastrointestinal)
Physical Findings
Ruddy cyanosis
Plethora
Hepatomegaly
Splenomegaly
Hypertension

PV occurs in two phases:
- Overt polycythemia characterized by a proliferation of erythroid cells
- Late phase of BM exhaustion referred to as post-polycythemia vera myelo-fibrosis (post-PV MF)

Rarely, a patient may present in a "pre-polycythemic" phase with decreased EPO and JAK2 mutation but without elevation of the Hgb or Hct. Evolution to overt polycythemia occurs eventually.

Clinical: Presenting symptoms and physical signs usually relate to thrombosis, hemorrhage and extramedullary hematopoiesis (Table 4.8).

Secondary and relative polycythemia: The diagnosis of PV requires that other secondary causes for polycythemia are ruled out. Testing for JAK2 plays an important role in ruling out secondary erythrocytosis since it is not present in these conditions. Secondary polycythemia is usually related to hypoxia but may also be due to truncated EPO receptors or abnormal EPO production by a tumor (renal cell carcinoma, hepatocellular carcinoma, cerebellar hemangioblastoma, parathyroid carcinoma). Relative (spurious or apparent) polycythemia is due to contraction of the plasma volume and is rare. Erythropoietin is sometimes administered as a doping agent to athletes in order to increase the oxygen-carrying capacity of the blood. Secondary polycythemia is usually easily differentiated from PV since these patients generally lack thrombocytosis and hepatosplenomegaly.

PB findings in overt polycythemia: The key PB and BM findings in the overt polycythemia phase are evident in the WHO bluebook criteria for PV (Table 4.9). The PB usually demonstrates an increase in RBCs, which are normochromic and normocytic. The presence of hemorrhage can modify this picture so that the hemoglobin (Hgb) and hematocrit (Hct) are normal, or even slightly decreased, with microcytosis and hypochromasia. Similarly, iron deficiency may result in normal

Table 4.9. WHO criteria for PV

Diagnosis requires both major criteria and one minor criterion or the first major criterion and two minor criteria.

Major Criteria

1. Hemoglobin >18.5 g/dL in men, 16.5 g/dL in women, or other evidence of increased RBC mass (elevated hematocrit or measured red cell mass)
2. Presence of the JAK2 V617F or JAK2 exon 12 mutation

Minor Criteria

1. BM biopsy showing hypercellularity with trilineage growth with prominent erythroid, granulocytic and megakaryocytic proliferation
2. Serum EPO level below the normal reference range
3. Endogenous erythroid colony formation in vitro

Adapted with permission from: Swerdlow SH, Campo E, Harris NL et al. World Health Organization Classification of Tumours of Haematopoietic and Lymphoid Tissues. Lyon: IARC, 2008:40, Table 2.02.

or low Hgb and Hct values. The WHO criteria assume adequate iron replacement in such cases. Approximately 50% of patients have elevated platelets, which, when accompanied by a normal hematocrit, can make differentiation of PV from ET quite difficult. A similar percentage of patients have neutrophilia, commonly with a slight left-shift, although blasts are rare in the PB. Basophilia is the norm, and eosinophilia is common. These findings differ from those observed in the PB of patients with secondary polycythemia in which the hematocrit rarely exceeds 60% and the platelets, WBCs, and basophils are not increased.

BM findings in overt polycythemia (Fig. 4.6): The BM in PV demonstrates panmyelosis; that is, increased normoblastic erythroid cells, myeloid cells and megakaryocytes. The megakaryocytes are characteristically prominent and are of various sizes with frequent clustering around sinuses and trabeculae but without significant atypia. The average cellularity of the BM is 80% and is almost always increased for the patient's age with only 10-15% being normocellular. Mild to moderate reticulin fibrosis may be present at initial BM examination. Stainable iron is not present. In secondary erythrocytosis, there is an increase in the erythroid cells only. The morphologic and clinical differences between PV and secondary erythrocytosis are summarized in Table 4.10.

Peripheral blood and BM in the post-PV MF phase: Progression of PV eventuates in post-PV MF, which is characterized by reticulin and sometimes collagen fibrosis of the BM. The PB demonstrates leukoerythroblastosis with teardrop cells. The erythrocytosis, thrombocytosis and neutrophilia seen in the overt phase now evolve to cytopenias. The BM, likewise, shows a decrease in the erythroid and myeloid cells, but the megakaryocytes usually remain prominent with increasing atypia. There may be a myeloid left-shift but with <10% blasts. No significant dysplasia is present. Dysplasia or >10% blasts signifies conversion to myelodysplasia, and ≥20% blasts to AML.

Erythropoietin levels play a critical role in ruling out PV in cases that do not demonstrate the JAK2 mutation. A small percentage of PV patients appear to be

Table 4.10. Clinical and morphological differences between PV and secondary erythrocytosis

PV	Secondary Erythrocytosis
JAK2 mutation present	JAK2 absent
Clinical symptoms related to hemorrhage and thrombosis	Clinical symptoms related to hypoxia, high oxygen affinity Hgb, or EPO producing tumor
Splenomegaly and/or hepatomegaly	Absent
50% have neutrophilia	Neutrophils normal in number
50% have thrombocytosis	Platelets normal in number
Granulocytic left-shift in PB common	Rare
Basophilia	Absent
Eosinophilia common	Absent
Hct typically >60%	Hct elevated but rarely >60%
Panmyelosis in BM	BM shows erythroid hyperplasia only
Prominent megakaryocytes with clustering	Megakaryocytes normal

negative for the JAK2 mutation, because they possess the exon 12 variant that is not detected by current commercially available testing methods. Therefore, negative PCR for JAK2 does not rule out PV. If JAK2 is negative and the EPO level is normal or elevated, then PV is ruled out.

Differential diagnosis of the post-PV MF phase: The post-PV MF phase of PV can be challenging or impossible to differentiate from the fibrotic phase of PM. When the early phase of the disease has not been observed, this distinction may not be possible. In such cases, the diagnosis "myeloproliferative neoplasm, unclassified" may be used. Because cytopenias occur in the post-PV MF phase of PV and the fibrotic phase of PM, differentiation of these from myelodysplasia with myelofibrosis may also be difficult. The BM is usually inaspirable in these diseases; therefore, the most distinguishing feature, dysplasia, may not be apparent. It is useful to do touch preparations (TP) of the fresh BM biopsy tissue when the marrow is found to be inaspirable. The Wright-Giemsa-stained TP slides yield cellular morphology comparable to that seen in an adequate BM aspirate, if cells are present. In some cases, the fibrosis has progressed to the point where even TPs are inadequate to evaluate for dysplasia. If no granulocytic dysplasia is present in the PB, then it may not be possible to rule out MDS. In these instances, a descriptive diagnosis may be given with an explanation and differential diagnosis discussed in the comment section of the report.

Genetics: The JAK2 mutation is present in virtually all cases of PV. The rare instances in which this abnormality is not identified are felt to be due either to the presence of the exon 12 variant of the JAK2 mutation, which is not detected by current methods outside of research applications, or a mutation burden that is too low for detection. The JAK2 mutation is not seen in reactive erythrocytosis or in MDS. Additional abnormalities may be seen on chromosomal analysis including +8, +9, del(20q), del(13q), and del(9p). These abnormalities increase with disease progression and are present in 80-90% of patients in the post-PV MF phase of disease and approximately 100% of patients with progression to MDS or AML. By definition, BCR/ABL is absent.

4

Red blood cell mass measurement and endogenous erythroid colonies (EECs): A few words should be said regarding two special tests that are referred to in the WHO PV criteria. Red blood cell mass is determined by taking an aliquot containing a known number of RBCs, labeling them with a radioactive tracer, injecting them back into the patient and measuring the dilution of the radioactivity. Demonstration of EECs requires taking BM cells from the patient, culturing them in vitro and observing whether of not erythroid colonies develop in the absence of stimulation by EPO, thus illustrating one of the classic characteristics of clonal neoplasms: the ability of the neoplastic clone to proliferate independent of mediators that normally regulate proliferation. These tests are usually not available in community hospitals and rarely done in large tertiary care centers.

Prognosis: Without treatment, the prognosis of PV is poor with a median survival of only a few months. Survival >10 years is common with treatment. Paradoxically, conversion to AML is 2-3% for untreated patients and up to 10% in patients with some kind of treatment. Most patients succumb due to thrombosis or hemorrhage, while approximately 20% die of AML or MDS. Clinical trials are currently ongoing using new agents specifically directed at the JAK2 mutation and show promise for the future.

Primary Myelofibrosis

Primary myelofibrosis is a clonal disorder of hematopoietic stem cells resulting in a proliferative process that predominately involves megakaryocytes and granulocytes. There is an associated reactive fibrosis (i.e., the fibroblasts are not part of the neoplastic clone). The megakaryocytes in these patients have decreased expression of the thromopoietin receptor (Mp1), which, paradoxically, results in increased sensitivity of the megakaryocytes to thrombopoietin leading to increased proliferation. Megakaryocytes and platelets release several cytotoxins that are known to stimulate fibrosis, including platelet-derived growth factor (PDGF), transforming growth factor beta (TGF-β), basic fibroblast growth factor (bFGF) and calmodulin. The older name for this disease is chronic idiopathic myelofibrosis (CIMF).

The worldwide incidence of PM ranges from about 2 to 13 cases per million persons per year, being approximately 8 to 10 cases/million/year in North America and Europe. The average patient is in the seventh decade of life and, although a wide range of ages have been reported, presentation before 40 years of age is rare. Most studies show a slight male predominance.

The natural history of PMF is that of an early **prefibrotic stage** that gradually evolves to a late **fibrotic stage**. Patients in the prefibrotic stage are usually asymptomatic and are discovered by unexpected CBC results or splenomegaly and/or hepatomegaly seen on a scan or, less commonly, palpated on physical examination.

Table 4.11. WHO criteria for the diagnosis of PM

Major Criteria
1. Presence of megakaryocyte proliferation and atypia, usually accompanied by reticulin and/or collagen fibrosis,

    or

    in the absence of significant reticulin fibrosis, the megakaryocyte changes must be accompanied by an increased BM cellularity characterized by granulocytic proliferation and often decreased erythropoiesis (i.e., prefibrotic cellular-phase disease)
2. Not meeting WHO criteria for PV, CML, MDS or other myeloid neoplasm
3. Demonstration of JAK2 or other clonal marker (e.g., MPL W515K/L),

    or

    in the absence of a clonal marker, no evidence that BM fibrosis is due to underlying inflammatory or other neoplastic diseases

Minor Criteria
1. Leukoerythroblastosis
2. Increase in serum lactate dehydrogenase level
3. Anemia
4. Splenomegaly

Diagnosis requires all three major criteria and two minor criteria

Adapted with permission from: Swerdlow SH, Campo E, Harris NL et al. World Health Organization Classification of Tumours of Haematopoietic and Lymphoid Tissues. Lyon: IARC, 2008:44, Table 2.04.

The majority, approximately 70%, are diagnosed in the fibrotic stage. The WHO criteria for PM are depictd in Table 4.11.

PB findings: The PB in the prefibrotic stage demonstrates the following:
- Mild to moderate anemia
- Leukocytosis
- Thrombocytosis

The elevated platelets can make differentiation from essential thrombocythemia (ET) and the overt phase of PV, when the hematocrit is decreased because of hemorrhage, difficult.

As PM gradually progresses to the fibrotic stage, the PB shows:
- Often, moderate to marked anemia
- Variable WBC and platelet counts
- Myelophthisis

Myelophthisis, the evidence of which is seen in the PB as leukoerythroblastosis with teardrop cells, is minimal or absent in the early phase of the disease and becomes more pronounced in proportion to the increase in BM fibrosis and extramedullary hematopoiesis (EMH).

BM findings in the prefibrotic stage:
- Hypercellularity due to panmyelosis and no or minimal reticulin fibrosis
- Megakaryocytes are often prominent and show clustering with abnormal nuclear lobation and bare nuclei.

- Ring sideroblasts are present in some cases of PM.

Again, differentiation from ET and PV may be difficult.

With disease progression, BM findings include the following:

- Bone marrow becomes less cellular
- Increasing prominence of the megakaryocytes
- Development of reticulin and collagen fibrosis

The increase in fibrosis results in dilation of the BM sinuses. The presence of hematopoiesis within the BM sinuses, sometimes referred to as intrasinusoidal EMH, is characteristic of PM.

Differential diagnosis from MDS with myelofibrosis and acute panmyelosis with myelofibrosis is accomplished by the presence of typical dysplastic changes in these entities. The megakaryocytes in MDS and acute panmyelosis with myelofibrosis lack clustering into large groups and are usually small and monolobated or hypolobated.

Conditions other than PM associated with myelofibrosis are summarized in Table 4.12.

Clinical findings in the early disease stage are minimal and may be related to thrombocytosis (thrombosis, bleeding) or anemia (fatigue). Scans may detect splenomegaly or hepatomegaly. In the fibrotic stage, hepatosplenomegaly due to EMH is typically massive. Extramedullary hematopoiesis may also occur in the LNs, dura matter, kidney, adrenal gland, gastrointestinal tract, lung and pleura, breast and skin. Other clinical findings become more common and more severe with disease progression and include fatigue, dyspnea, weight loss, night sweats, low-grade fever, gouty arthritis and renal stones.

A comparison of the clinical and morphological findings in the prefibrotic and fibrotic stages of PM is given in Table 4.13.

Table 4.12. Conditions associated with myelofibrosis

Other MPNs:
    ET
    PV
    CML
Hodgkin and non-Hodgkin lymphomas
Other hematologic malignancies:
    AML
    ALL
    HCL
    PCM
Metastatic carcinoma:
    Breast
    Prostate
Tuberculosis and other infections
Toxins:
    Radiation
    Benzene

HCL: hairy cell leukemia; PCM: plasma cell myeloma

Table 4.13. Clinical and morphological findings in the prefibrotic and fibrotic phases of PM

Prefibrotic	Fibrotic
No or minimal splenomegaly/ hepatomegaly	Moderate to massive splenomegaly/ hepatomegaly
Mild anemia	Moderate to marked anemia
Leukocytosis	WBC variable
Thrombocytosis	Platelet count variable
Leukoerythroblastosis mild or absent	Leukoerythroblastosis present
RBC poikilocytosis minimal or absent	Prominent RBC poikilocytosis
Teardrop cells few or absent	Teardrop cells present
Minimal or no reticulin fibrosis of BM	Reticulin and/or collagen fibrosis present
BM panmyelosis with prominent megakaryocytes with atypical features	Megakaryocytes with atypical features but decreased BM cellularity
BM sinuses normal	BM sinuses dilated, some with EMH
Osteosclerosis absent	Osteosclerosis (new bone formation) present

Genetics: There is no specific genetic lesion for PM, but cytogenetic abnormalities do occur in about 60% of cases including del(13q), del(20q), partial trisomy 1q, +8 and +9. Deletions of chromosomes 7 and 5 occur in relation to treatment of PM with cytotoxic therapy. By definition, BCR/ABL is absent. JAK2 is detected in 43-52% of patients with PM. Up to 5% of cases possess the MPL W515K/L mutation, which is also sometimes seen in ET.

Prognosis: Survival times for PM range from months to decades with a median of approximately five years. Progression to AML occurs in 5-30% and is a part of the natural history of PM. Poor prognostic factors include the following:

- Age >70 years
- Hemoglobin <10 g/dL
- Platelet count <100 x 10^9/L
- Marked granulocytic left-shift in the PB
- Presence of a genetic abnormality

The major complications leading to death include BM failure, thrombosis, portal hypertension and cardiac failure.

Essential Thrombocythemia

Essential thrombocythemia is a clonal disorder of the BM stem cells; however, there are cases that fit the definition of ET in which clonality cannot be demonstrated. The incidence is about 1.5 cases per 100,000 in population per year. Most cases are identified in the sixth decade of life with a second smaller peak

Table 4.14. Thrombotic and hemorrhagic events in ET

Microvascular Thrombosis	Large Vessel Thrombosis	Hemorrhage
Headache	Stroke	Ecchymosis
Transient ischemic attacks	Myocardial infarction	Epistaxis
Dizziness	Deep vein thrombosis	Gastrointestinal bleeding
Visual disturbances	Splenic vein thrombosis	Genitourinary bleeding
Digital pain or gangrene	Hepatic vein thrombosis	Bleeding in upper airway passages

occurring in patients at approximately 30 years of age, at which time there is a slight female predominance. Essential thrombocythemia can occur in children, but this is rare.

Clinical: Approximately one-third of patients are asymptomatic at the time of diagnosis, being discovered incidentally by blood testing for other reasons. The clinical findings are usually related to thrombosis or hemorrhage (Table 4.14). Thrombosis with ET occurs more commonly in the arterial rather than the venous system. Fifty percent of patients have splenomegaly and 20% have hepatomegaly, neither of which is usually massive.

PB findings: The PB shows increased platelets ≥450 x 10⁹/L and frequently >1000 x 10⁹/L. The platelets may show abnormalities of morphology or granulation, but this is rare. Basophilia, leukoerythroblastosis and teardrop cells should be absent and, if present, are indicative of other MPNs. The WBC is usually normal, and the RBCs are normocytic and normochromic but may be microcytic and hypochromic if significant hemorrhage (Fe deficiency) has occurred.

BM findings: The BM is normocellular or mildly hypercellular, with rare cases being hypocellular. The most important feature is the presence of increased numbers of megakaryocytes that are characteristically large or giant with loose clustering and hyperlobation. Bizarre megakaryocytes like those occurring in PM are not seen. The granulocytic cells are normal or slightly increased. The erythroid cells are normal but may be increased in patients that have hemorrhage.

Table 4.15. Causes of reactive thrombocytosis

- Iron deficiency
- Splenectomy
- Surgery
- Infection
- Inflammation
- Connective tissue diseases
- Metastatic cancer
- Lymphoproliferative disorders

Table 4.16. WHO criteria for essential thrombocytosis

1. Sustained platelet count ≥450 x 10⁹/L
2. BM biopsy showing proliferation mainly of the megakaryocytic lineage with increased numbers of enlarged, mature megakaryocytes; no increase or left-shift of neutrophil granulopoiesis or erythropoiesis
3. Not meeting WHO criteria for PV, PM, CML, MDS or other myeloid neoplasm
4. Demonstration of JAK2 or other clonal marker, or in the absence of a clonal marker, no evidence for reactive thrombocytosis*

*Presence of a condition associated with reactive thrombocytosis does not rule out ET if the first three criteria are met. Adapted with permission from: Swerdlow SH, Campo E, Harris NL et al. World Health Organization Classification of Tumours of Haematopoietic and Lymphoid Tissues. Lyon: IARC, 2008:48, Table 2.06.

Secondary thrombocytosis due to inflammatory or other reactive causes, or associated with non-hematologic neoplasms, must be ruled out. Causes of reactive thrombocytosis are listed in Table 4.15. In most, the clinical picture will be dominated by the features of the primary disease rather than the thrombotic and hemorrhagic features due to platelet dysfunction seen with ET. Hemorrhage and thrombosis are rare with secondary thrombocytosis. Toxic changes in the neutrophils, seen on PB examination, may be a clue for an infectious etiology. The mechanism for thrombocytosis in inflammatory conditions is the release of cytokines, the most important of which is interleukin-6 (IL-6), that stimulate megakaryocytic proliferation. Interleukin-6 also stimulates C-reactive protein (CRP) production. In contrast to IL-6, CRP measurement is inexpensive and available in most laboratories and may be used as a surrogate test for IL-6, which, if elevated, favors a reactive etiology for the increased platelets. A normal serum ferritin rules out iron deficiency as a cause for the thrombocytosis. The dogma that ferritin is an acute phase protein that is elevated in reactive conditions is overemphasized since this elevation is minimal. Significant hemorrhage with ET resulting in iron deficiency makes differentiation from secondary thrombocytosis difficult. In these cases, other clinical features such as an obvious cause for the hemorrhage, other than platelet dysfunction, must be relied upon. The JAK2 mutation is not present in secondary thrombocytosis. The WHO criteria for the diagnosis of ET are given in Table 4.16.

Genetics: Not only is there no specific genetic marker for ET, genetic abnormalities are rare, occurring in only 5-10% of patients. Genetic defects that have been observed in ET include +8, 9q abnormalities and del(20q). By definition, BCR/ABL and the MDS associated abnormalities (see chapter on MDS) must be absent. The JAK2 mutation is seen in 41-57% and MPL W515K/L in about 1%, neither of which is seen with reactive thrombocytosis.

Prognosis: The clinical course of ET is usually indolent and commonly does not affect the life expectancy of the typical elderly patient. Many patients require only aspirin, but those with significant hemorrhagic and thrombotic complications may require myelosuppressive therapy. Acute myeloid leukemia develops in <5% of cases and is frequently related to previous cytotoxic therapy.

Table 4.17. Classification of blood eosinophilia

```
                        ┌─────────────────────┐
                        │  Blood Eosinophilia │
                        └─────────────────────┘
                          ╱                 ╲
                 ┌──────────────┐      ┌──────────────┐
                 │   Acquired   │      │   Familial   │
                 └──────────────┘      └──────────────┘
                  ╱         ╲
         ┌──────────────┐        ┌──────────────┐
         │   Primary    │        │  Secondary   │
         └──────────────┘        └──────────────┘
          ╱        ╲              ╱            ╲
```

| Clonal:
ALL
AML
MPN
MDS
MDS/MPN
PDGFRA
PGDFRB
FGFR1
CEL | Idiopathic
(HES) | Infectious:

Tissue-
invasive
parasites
Rarely
bacteria
or virus | Non-
infectious:

Drugs
Toxins
Allergies
Autoimmune
disorders
Malignacies
Endocrine
diseases |

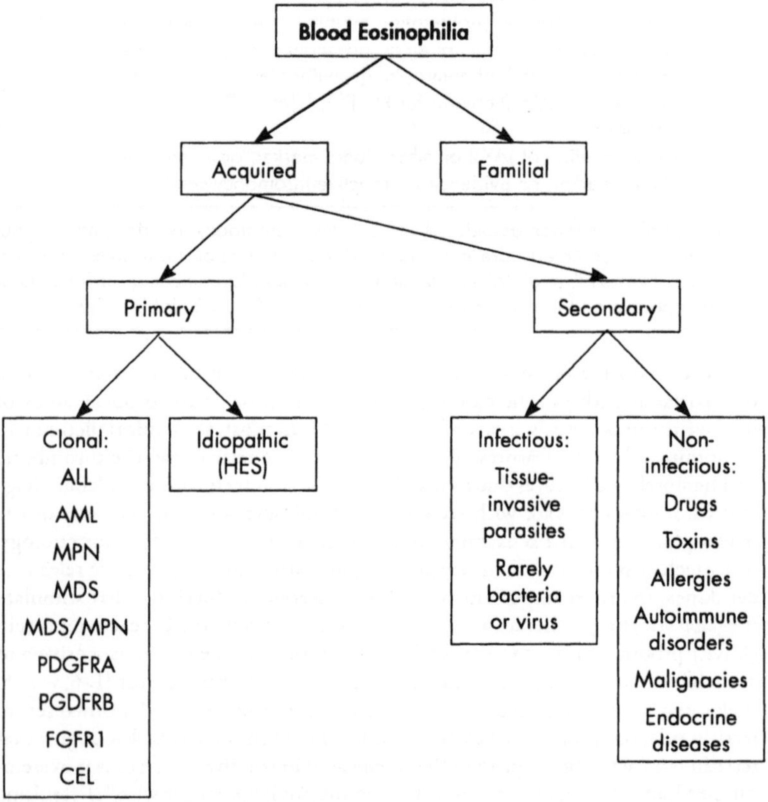

HES: hypereosinophilic syndrome

Chronic Eosinophilic Leukemia, Not Otherwise Categorized

The classification of blood eosinophilia is depicted in Table 4.17. From this table it can be seen that clonal eosinophilia may accompany a variety of hematologic neoplasms. Eosinophilia may also be associated with myeloid neoplasms in which there are chromosomal anomalies of PDGFRA, PDGFRB or FGFR1. These cases were frequently diagnosed as chronic eosinophilic leukemia (CEL) or systemic mastocytosis (SM) under the original WHO classification scheme. However, modern research techniques have subsequently defined these cases as distinct entities with important clinical implications that justify their own category in the revised WHO scheme. Cases of clonal eosinophilia that are not associated with chromosomal abnormalities of PDGFRA/PDGFRB/FGFR1 and that do

Table 4.18. WHO criteria for CEL

1. Eosinophilia ≥1.5 x 10⁹/L
2. No BCR/ABL or other MPN (PV, ET, PM) or MDS/MPN
3. No t(5;12)(q31-35;p13) or other rearrangement of PDGFRB
4. No FIP1L1-PDGFRA or other rearrangement of PDGFRA
5. No rearrangement of FGFR1
6. Blast count in the PB and BM <20% and no inv(16)(p13.1;q22) or t(16;16)(p13.1;q22) or other feature diagnostic of AML
7. There is a clonal cytogenetic or molecular genetic abnormality, or blasts are >2% in the PB or >5% in the BM

Adapted with permission from: Swerdlow SH, Campo E, Harris NL et al. World Health Organization Classification of Tumours of Haematopoietic and Lymphoid Tissues. Lyon: IARC, 2008:52, Table 2.08.

not accompany other defined hematologic neoplasms are placed into CEL, not otherwise categorized (NOC).

Definition: The definition of CEL requires the exclusion of non-neoplastic and neoplastic disorders that cause reactive eosinophilia, other neoplasms in which eosinophilia is part of the neoplastic clone, and a T-cell population with abnormal cytokine production stimulating eosinophil proliferation. This is primarily accomplished by observing the absence of morphologic features of these disorders in the BM and PB in a setting of ≥1.5 x 10⁹/L PB eosinophils, usually with organ damage. The WHO criteria for CEL are given in Table 4.18.

The true incidence of CEL is difficult to determine because of the difficulty of differentiating CEL from hypereosinophilic syndrome (HES) and the fact that some cases that were diagnosed as CEL in the past would now be placed into newly recognized disorders. The peak incidence is in the fourth decade of life, and there is a marked male predominance.

Clinical: Regardless of the cause, moderate (1.5 to 5 x 10⁹/L) to severe (>5 x 10⁹/L) eosinophilia can result in target organ damage that is mediated by the release of cytokines from eosinophilic granules including major basic protein, eosinophil cationic protein, and eosinophil-derived neurotoxin. The major organs which can be affected include the following:

• Heart
• Lungs
• Skin
• Nervous system
• Vasculature
• Gastrointestinal tract

Target organ damage is rare with mild (0.6 to 1.5 x 10⁹/L) eosinophilia. Other symptoms can be due to the BM hyperproliferative state, such as fever, fatigue and weight loss. Approximately 10% of patients are discovered incidentally on blood counts done for other reasons and are clinically asymptomatic.

PB findings: The PB eosinophils may demonstrate a variety of morphologic abnormalities including the following:
- Hypogranulation
- Clear areas within the cytoplasm
- Cytoplasmic vacuolization
- Hypo- or hypersegmentation
- Cytomegaly

These abnormalities are nonspecific and can be seen in reactive eosinophilia as well as CEL and HES. Myeloblasts are absent or are seen in small numbers in the PB. Some cases are accompanied by neutrophilia, monocytosis or mild basophilia.

BM findings: The BM is hypercellular with a predominance of eosinophils that usually demonstrate a relatively normal maturation sequence. The erythroid and megakaryocytic cells are usually normal. There may be dysplastic changes in the other cell lines and/or increased myeloblasts, but these features, although helpful when present, are not specific for CEL. Bone marrow fibrosis is seen in some cases. Bone marrow evidence of other disorders which are associated with eosinophilia should be absent.

Genetics: No genetic abnormality is specific to CEL. Cytogenetics and special studies must be done to rule out BCR/ABL and abnormalities of PDGFRA, PDGFRB, and FGFR1. Clonality in CEL is difficult to demonstrate and is usually seen in the form of myeloid associated karyotypic abnormalities such as +8 and i(17q).

Prognosis: The prognosis for CEL is variable, but the five-year survival approaches 80%. Unfavorable prognostic findings include the following:
- Increased myeloblasts in the PB or BM
- Dysplasia in other cell lines
- Marked splenomegaly

Mastocytosis

Mast cell disease includes a diverse group of disorders ranging from a clinically mild skin rash to aggressive, deadly systemic diseases. In common among them is the abnormal proliferation and accumulation of mast cells which are derived from hematologic stem cells. For this reason, mast cell disease is now logically placed with myeloproliferative neoplasms in the revised WHO classification scheme. Some of the disorders which were previously diagnosed as systemic mastocytosis (SM) or CEL by the criteria in the original WHO bluebook have subsequently been characterized as clonal disorders with clinical features that justify their own category (myeloid neoplasms associated with eosinophilia with abnormalities of PDGFRA, PDGFRB, FGFR1). The remaining mast cell disorders are categorized by their morphologic and clinical features into cutaneous mastocytosis (CM) and SM.

Cutaneous mastocytosis is infiltration of the skin by mast cells without evidence of systemic involvement. Elevated total serum tryptase (>20 ng/mL) or organomegaly should prompt appropriate investigation for SM. Approximately 80% of patients have skin involvement only. Cutaneous mastocytosis is subclassified into urticaria pigmentosa, diffuse cutaneous mastocytosis, and solitary mastocytoma of skin. By definition, the BM is not involved in CM. Approximately 50% of patients with SM have skin involvement.

Table 4.19. Signs and symptoms of SM

Constitutional Symptoms Related to Hyperproliferation
Fatigue
Weight loss
Fever
Sweats

Skin Manifestations
Pruritis
Urticaria
Dermatographism

Mediator Related
Abdominal pain and GI distress
Flushing
Syncope
Hypertension
Headache
Hypotension
Tachycardia
Asthma-like symptoms

Bone Related
Bone pain
Joint pain
Fractures

Systemic mastocytosis comprises approximately 20% of cases of mastocytosis. Systemic mastocytosis typically occurs after the third decade of life, and some studies show a female predominance. Any organ may be involved, but frequent sites include the skin, spleen, lymph nodes, liver and gastrointestinal tract. Clinical symptoms may be due to the release of biochemical mediators, such as histamine, eicosanoids, proteases or heparin (see Table 4.19). Common physical findings include splenomegaly, lymphadenopathy and, less frequently, hepatomegaly.

PB findings: Mast cells are rarely observed in the PB, but other hematologic abnormalities are common, including anemia, leukocytosis or leukopenia, and thrombocytosis or thrombocytopenia. Extensive infiltration of the BM can result in BM failure and pancytopenia. Total serum tryptase with levels >20 ng/mL are indicative of SM, while lesser elevations are seen with CM.

Bone marrow biopsy findings and IHC: Bone marrow is the specimen in which SM is most commonly identified, usually the biopsy. The BM biopsy demonstrates sharply demarcated or loose aggregates of mast cells distributed in a paratrabecular or perivascular pattern. Classically, there is a core of mature lymphocytes surrounded by mast cells with an outer cuff of eosinophils. However, this is not always the case, and loose naked groups of mast cells may be seen. On hemotoxylin- and eosin-stained sections, mast cells have eccentrically placed nuclei and abundant eosinophilic cytoplasm containing faint granules. Associated fibrosis and thickening of adjacent bony trabeculae and/or reticulin or collagen fibrosis of the marrow are common. In some cases, the mast cell groups are quite subtle and differentiation from fibrosis or normal connective tissue requires IHC. The classic stain for mast cells is toluidine blue, but decalcification of the BM biopsy tissue decreases reactivity. Tryptase IHC is positive in all mast cells, while only a subpopulation is positive for chymase. Mast cells are also positive for CD45(LCA), CD68 and CD117. Neoplastic mast cells are positive for CD2 and CD25, which is lacking in normal mast cells. Occasionally, the mast cells have clear cytoplasm resembling hairy cell leukemia in BM sections but are easily distinguished by the absence of positivity for CD20.

BM aspirate findings: Mast cells are difficult to aspirate because of their paratrabecular location and associated fibrosis. When present in the aspirate, they usually occur in groups with reniform nuclei and pale hypogranular cytoplasm. Less commonly, their appearance is similar to that of normal mast cells. Cytochemically, mast cells are positive for chloracetate esterase (CAE) and are negative for myeloperoxidase (MPO).

Reactive mast cells: Non-neoplastic mast cells may be seen in increased numbers in the BM as a reactive phenomenon in various conditions including other hematologic neoplasms. Non-clonal mast cells usually lack atypia and are arranged interstitially.

Concurrent hematologic neoplasms: Approximately 20% of cases of SM have concurrent involvement by another hematologic neoplasm, which should be classified according to applicable WHO criteria. Therefore, it is extremely important to carefully evaluate the BM for other disorders such as AML, MDS, other MPNs and lymphoma.

Mast cell leukemia is a rare, aggressive disorder in which mast cells are seen in the PB, diffusely involve the BM biopsy specimen and are seen in large sheets in the BM aspirate slides.

The WHO criteria for SM are given in Table 4.20.

Flow cytometry: By FCM normal mast cells express CD45, CD33, and CD117 and are negative for CD25, CD14, CD15, CD16 and CD2. Neoplastic mast cells

---

Table 4.20. BM criteria for SM diagnosis requires the presence of the major and one minor, or three minor criteria

Major Criteria
- Multifocal infiltrates of mast cells (15 or more cells in aggregates) of BM and/or other extracutaneous organs

Minor Criteria
- In BM or other extracutaneous tissue sections, more than 25% of the mast cells are spindled or morphologically atypical, or, more than 25% of the mast cells in aspirate smears are immature or atypical.
- Detection of *KIT* point mutation at codon 816 in BM, PB, or other extracutaneous tissue
- Mast cells co-express CD2 with CD25
- Total serum tryptase >20 ng/mL (unless there is an associated clonal myeloid disorder)

Adapted with permission from: Swerdlow SH, Campo E, Harris NL et al. World Health Organization Classification of Tumours of Haematopoietic and Lymphoid Tissues. Lyon: IARC, 2008:56, Table 2.10.

Table 4.21. WHO criteria for SM variants

ISM:
- Meets criteria for SM
- No "C" findings
- No evidence of associated hematologic neoplasms
- Mast cell burden is low with invariable skin involvement

    BM mastocytosis: as above with BM involvement and no skin involvement
    Smoldering SM: as above but with two or more "B" findings but no "C" findings

SM-AHNMD:
- Meets criteria for SM and
- Associated non-mast cell hematologic neoplasms that meet WHO criteria for distinct entity.

ASM:
- Meets criteria for SM
- One or more "C" findings
- No evidence of mast cell leukemia
- Usually no skin lesions

    Lymphadenopathic mastocytosis: progressive lymphadenopathy with PB eosinophilia, often with extensive bony involvement, and hepatosplenomegaly, usually without skin involvement

Mast Cell Leukemia:
- Meets criteria for SM
- Biopsy shows diffuse infiltration by atypical, immature mast cells
- Aspirate shows ≥20% mast cells
- PB shows ≥10% mast cells

    Aleukemic mast cell leukemia: rare variant; as above but <10% mast cells in PB

"B" Findings
1. BM showing >30% infiltration by mast cells and/or serum total tryptase >200 ng/mL
2. Signs of dysplasia or myeloproliferation in non-mast cell lineage, but without criteria to be distinct WHO entity
3. Hepatomegaly without incidence of liver dysfunction, and/or palpable splenomegaly without hypersplenism, and/or lymphadenopathy by palpation or imaging

"C" Findings
1. BM dysfunction manifested by one or more cytopenias (ANC <1.0 x 10^9/L, Hgb <10 g/dL or plt <100 x 10^9/L) but no frank non-mast cell hematologic neoplasm
2. Hepatomegaly with liver dysfunction, ascites, and/or portal hypertension
3. Skeletal involvement with large osteolysis and/or pathological fractures
4. Palpable splenomegaly with hypersplenism
5. Malabsorption with weight loss due to GI infiltration by mast cells

Adapted with permission from: Swerdlow SH, Campo E, Harris NL et al. World Health Organization Classification of Tumours of Haematopoietic and Lymphoid Tissues. Lyon: IARC, 2008:57, Table 2.12.

Table 4.22. WHO criteria for mast cell sarcoma and extracutaneous mastocytoma

Mast Cell Sarcoma
- Unifocal mast cell tumor
- No evidence of SM
- No skin lesions
- Destructive growth
- High-grade cytology

Extracutaneous Mastocytoma
- Unifocal mast cell tumor
- No evidence of SM
- No skin lesions
- Non-destructive growth
- Low-grade cytology

Adapted with permission from: Swerdlow SH, Campo E, Harris NL et al. World Health Organization Classification of Tumours of Haematopoietic and Lymphoid Tissues. Lyon: IARC, 2008:57, Table 2.12.

are additionally positive for CD2, CD25 and CD35. Mast cells are distinguished on FCM by bright positivity for CD117 and surface IgE.

Systemic mastocytosis variants: The WHO bluebook defines four variants of SM: indolent SM (ISM), SM with associated clonal hematological non-mast cell lineage disease (SM-AHNMD), aggressive SM (ASM) and mast cell leukemia. The morphology of mast cell leukemia and SM-AHNMD are discussed above. The WHO criteria for these variants are given in Table 4.21.

Prognosis: The prognosis of SM depends on the variant. Indolent SM has little or no affect on life expectancy, while patients with ASM or mast cell leukemia may survive for only weeks to months. There is currently no cure for SM, although BM transplantation is being tried in some advanced cases. In contrast, imatinib treatment is effective in many cases of myeloid neoplasms associated with eosinophilia with abnormalities of PDGFRA, PDGFRB or FGFR1. The prognosis for SM-AHNMD is usually that of the associated neoplasm.

Mast cell sarcoma (MCS) and extracutaneous mastocytoma: Mast cell sarcoma is a very rare tumor with destructive growth composed of highly atypical mast cells. This tumor may show distant spread or a subsequent leukemic phase.

In contrast, extracutaneous mastocytoma is a localized collection of mature mast cells. Most cases of this very rare disorder have been localized to the lung. The WHO criteria for mast cell sarcoma and extracutaneous mastocytoma are given in Table 4.22.

Myeloproliferative Neoplasms, Unclassifiable

Common situations in which MPNs are not classifiable: This diagnosis, MPN, U, is used in cases where there is strong evidence of a clonal proliferative disease that does not fit the criteria for one of the other MPNs. Many cases are early stages of a MPN (PV, PM, ET) that have not fully developed the features to differentiate them from one another. Others represent a late fibrotic phase MPN in which the features of the original MPN have been obscured and cases with overlapping features of multiple MPNs.

In the first situation, a non-fully developed MPN, follow-up studies within six months will usually yield findings that enable classification.

In instances of late fibrotic disease, a precise diagnosis may not be possible. Therefore, any features which could be helpful to the clinician should be emphasized in a descriptive diagnosis.

In the case of overlapping features with other MPNs, careful evaluation of the clinical, laboratory and morphologic findings will usually resolve the diagnostic confusion.

MPNs in which important information is missing: The diagnosis of MPN, U should not be used when insufficient information is available for classification. The most common situation is when the results of genetic studies are not available. The best way to deal with these cases is to state that the diagnosis is presumed and that the additional studies should be obtained or are pending.

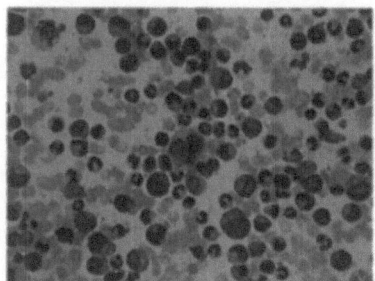

Figure 4.1. CML, BM aspirate showing granulocytic hyperplasia with full range of maturation (200x).

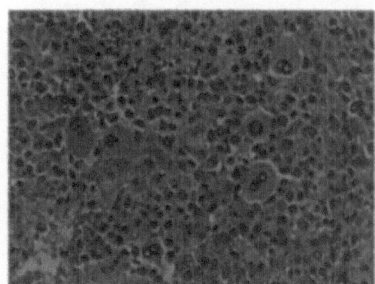

Figure 4.2. CML, BM biopsy with hyperplasia consisting of granulocytes and small megakaryocytes (100x).

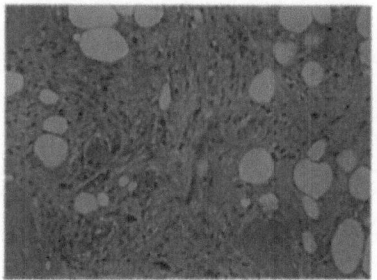

Figure 4.3. PM, BM biopsy (100x).

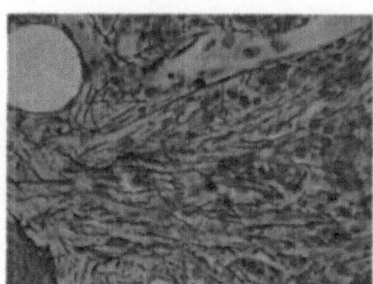

Figure 4.4. PM, reticulin stain of BM biopsy showing marked fibrosis (100x).

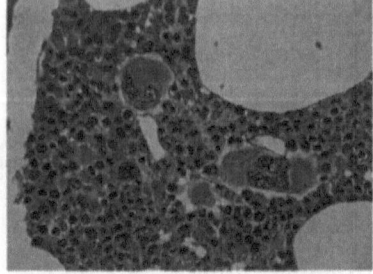

Figure 4.5. ET, BM biopsy showing increased large megakaryocytes (100x); compare to megakaryocytes seen with CML in Figure 4.2.

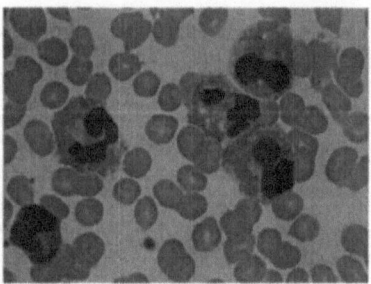

Figure 4.6. CEL, abnormal eosinophils in PB (500x).

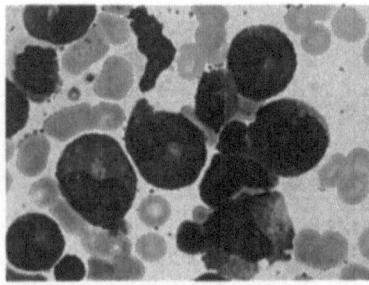

Figure 4.7. CEL, BM aspirate with increased eosinophilic precursor cells and slight atypia (500x).

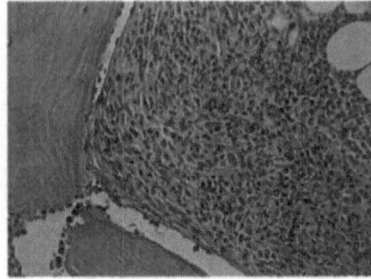

Figure 4.8. SM, BM biopsy showing paratrabecular aggregate of neoplastic mast cells with intermixed eosinophils (200x).

CHAPTER 5

Myelodysplastic/Myeloproliferative Neoplasms

Introduction

Effective versus ineffective hematopoiesis: As previously discussed, myelodysplastic syndromes (MDSs) are clonal disorders of hematopoietic cells resulting in ineffective hyperproliferation. That is, hypercellularity of the bone marrow (BM) with morphologic and functional defects of the affected cell line(s), and peripheral blood (PB) cytopenias. Chronic myeloproliferative neoplasms (MPNs), on the other hand, are clonal disorders of hematopoietic cells resulting in effective hyperproliferation. That is, BM hypercellularity with relatively normal cell morphology and function, and elevated PB cell counts. Some patients, however, defy this classification scheme and demonstrate overlapping features of both MDSs and MPNs.

Rationale for the myelodysplastic/myeloproliferative neoplasm (MDS/MPN) group: The French American British (FAB) classification system insisted on squeezing these disorders into either MDS or MPN, based on the predominant features even in cases that possess glaring evidence of both. Despite great strides in research methods, the fundamental pathogenic mechanisms of MDS, MDS/MPN and most MPNs (an exception is CML) are not fully understood and may or may not have commonalities. The current WHO classification recognizes this lack of scientific rationale for assigning these neoplasms into either the MDS or MPN category based on current understanding of the pathology or clinical observations.

Chronic Myelomonocytic Leukemia, Not Otherwise Categorized

Chronic myelomonocytic leukemia is a clonal hematopoietic stem cell neoplasm characterized by persistent monocytosis and the presence of dysplasia in one or more myeloid cell lines.

Chronic myelomonocytic leukemia is the prototypic and most common of the MDS/MPNs, but the exact incidence is difficult to determine since it has historically been placed with MDS in some studies and MPN in others. The occurrence can be roughly estimated from these studies to be 4 cases per 100,000 in population per year. The median age at presentation is 65 to 75 years with a male to female ratio of approximately 2:1.

Clinical: The presenting symptoms may be related to cytopenias and/or hyperproliferation and commonly include fatigue, fever, weight loss and night sweats. Splenomegaly is present in approximately 40% of patients and hepatomegaly in 20%, regardless of whether dysplasia or proliferation is predominant, but are more common when the WBC count is elevated. Lymphadenopathy may also occur.

Bone Marrow: A Practical Manual, by Daniel A. Cherry and Tomislav M. Jelic.
©2011 Landes Bioscience.

Peripheral blood findings: By definition, the absolute monocyte count in the PB must be >1 x 10⁹/L, and they usually make up >10% of the differential count. The monocytes are mostly mature with normal morphology but may show hyperlobation, increased cytoplasmic basophilia and/or abnormal granulation. Dysplasia may also be seen in the neutrophils, RBCs or platelets (see discussion of the morphologic features of dysplasia in MDS chapter). The general pattern of dysplasia, cell counts and clinical symptoms is logical. The cell lines with prominent dysplasia tend to be decreased (cytopenias) and dysfunctional with related clinical findings (bleeding, anemia, infections). The cell lines in which proliferation predominates (elevated cell counts) tend to be more normal, both morphologically and functionally, with clinical features related to hyperproliferation, such as night sweats, weight loss and organomegaly. The PB findings vary according to a mixture of these features. A slight granulocytic left-shift may be present, but myelocytes and promyelocytes usually make up <10% of the differential WBC count. Blasts and promonocytes, which are considered as blast equivalents in CMML by definition, account for fewer than the 20% that would justify a diagnosis of acute myeloid leukemia (AML). Mild normocytic or microcytic anemia and moderate thrombocytopenia are common. Large atypical platelets may be seen.

Bone marrow findings: The BM is hypercellular in approximately 75% of patients but may be normocellular or hypocellular. Monocytic proliferation is always present, granulocytic proliferation is common and erythroid proliferation occurs sometimes. Monocytes are notoriously difficult to identify morphologically in the BM; therefore, cytochemistry using nonspecific esterase (NSE) and/or immunohistochemistry (CD68, CD163) should be routinely employed if CMML is suspected. Granulocytic dysplasia is present in most cases, megakaryocytic dysplasia in approximately 80% and erythroid dysplasia in 50%. Reticulin fibrosis can be identified by a special stain in about 30%. The WHO criteria for the diagnosis of CMML are based mostly on morphologic findings (Table 5.1). Myeloblasts, monoblasts and promonocytes are all blast equivalents for classification purposes with CMML. Promonocytes have large amounts of gray to basophilic cytoplasm, occasional purple cytoplasmic granules, delicate nuclear folding or creases and variable prominent nucleoli.

Table 5.1. WHO criteria for CMML

1. Persistent PB monocytosis >1.0 x 10⁹/L
2. Negative for Philadelphia chromosome
3. Fewer that 20% blast equivalents (myeloblasts + promonocytes) in PB and BM
4. Dysplasia in one or more myeloid cell lines. If dysplasia is minimal or absent, CMML may be diagnosed if the other criteria are present and the following are true:
   - an acquired, clonal cytogenetic abnormality is present in the BM cells, or
   - the monocytosis has been present for at least three months and
   - all other causes for the monocytosis are ruled out.

Adapted with permission from: Swerdlow SH, Campo E, Harris NL et al. World Health Organization Classification of Tumours of Haematopoietic and Lymphoid Tissues. Lyon: IARC, 2008:76, Table 4.01.

Table 5.2. WHO defined subgroups of CMML

CMML-1	Blasts <5% in PB, <10% in BM
CMML-2	Blasts 5-19% in PB, or 10-19% in BM, or when Auer rods are present.

Subgroups of CMML: The WHO bluebook divides CMML into two subgroups based on the blast count and the presence of Auer rods, in a fashion similar to that of the MDS, refractory anemia with excess blasts. Table 5.2 summarizes the definitions of these subgroups. The previously described subgroup of CMML with eosinophilia is no longer placed in the MDS/MPN group and, instead, has been placed into the newly defined category, myeloid neoplasms associated with eosinophilia with abnormalities of PDGFRA, PDGFRB or FGFR1. This disorder is rare.

Genetics: Approximately one-third of cases have a BM chromosomal abnormality, most commonly +8, -7/del(7q), and structural abnormalities of 12p. By definition, BCR/ABL is not present. Isolated i(17q) is associated with marked granulocytic dysplasia (hyposegmentation) and a high risk for transformation to AML. Some experts believe these cases should be a unique entity. Abnormalities of 11q23 are rare in CMML and suggest AML.

Prognosis: The median survival with CMML is 30 months, with a range of 1 to 100 months. The percentage of blasts in the BM and PB is the most important predictor of survival. Rate of progression to AML is 15-30%.

Atypical Chronic Myeloid Leukemia, Not Otherwise Categorized

Atypical chronic myeloid leukemia (aCML) is a clonal BM stem cell neoplasm characterized by neutrophilia and neutrophil dysplasia. This is a disease of the elderly with an approximate incidence of slightly less than 1 per 100,000 in population annually and a male-to-female ratio of about 2:1.

Clinical findings: The clinical features are related to cytopenias, usually anemia and, less commonly, thrombocytopenia. The spleen may be involved by disease with clinical symptoms of splenomegaly (i.e., tugging or full sensation in upper abdomen).

Peripheral blood findings: The PB demonstrates neutrophilia which can be >300 x 10^9/L with dysgranulocytopoiesis (see discussion of dysplastic morphology in MDS chapter) predominately in the form of pseudo-Pelger-Huët change. Multilineage dysplasia may be present. The granulocytes are left-shifted with PB promyelocytes, myelocytes and metamyelocytes comprising ≥10% of the differential cell count. Blasts are <20%, by definition. Monocytosis and/or basophilia may occur but are not prominent. Anemia and thrombocytopenia are common. The leukocyte acid phosphatase (LAP) score is variable and not useful for diagnosis.

Bone marrow findings: The BM is hypercellular due to the granulocytic hyperproliferation. A granulocytic left-shift is usually present, but blasts are <20% and do not occur in large groups or sheets as in AML. Granulocytic dysplasia is present by definition. The erythroid and megakaryocytic cells are variable in number and may be dysplastic. Some cases demonstrate reticulin fibrosis. The WHO diagnostic criteria are given in Table 5.3.

Table 5.3. WHO diagnostic criteria for aCML

- PB leukocytosis (WBC ≥13 x 10⁹/L) due to increased mature and immature granulocytes with prominent granulocytic dysplasia
- No BCR/ABL
- Neutrophil precursors (promyelocytes, myelocytes, and metamyelocytes) ≥10% of WBCs
- No or minimal absolute basophilia; basophils <2% of WBCs
- No or minimal monocytosis (<10% of leukocytes)
- BM hypercellularity due to increased granulocytes with granulocytic dysplasia with or without erythroid and/or megakaryocytic dysplasia
- Blast <20% in BM and PB

Adapted with permission from: Swerdlow SH, Campo E, Harris NL et al. World Health Organization Classification of Tumours of Haematopoietic and Lymphoid Tissues. Lyon: IARC, 2008:80, Table 4.02.

Syndrome of abnormal chromatin clumping: The WHO classification scheme recognizes this variant of aCML which is characterized by prominent clumping of the granulocytic nuclear chromatin, usually accompanied by hypolobation and hypogranulation. Erythroid and megakaryocytic dysplasia are common, but severe anemia and/or thrombocytopenia are uncommon. The WBC count is usually elevated but may be normal. The prognosis is similar to that of non-variant aCML.

Genetics: Genetic abnormalities are found in about 80% of cases and include defects commonly seen with myeloid neoplasms such as +8, +13, del(20q), i(17q) and del(12p). None are specific for aCML. By definition, BCR/ABL is absent. A few cases with abnormal PDGFRB were previously diagnosed as aCML and are now in the new WHO category of myeloid neoplasms associated with eosinophilia with abnormalities of PDGFRA, PDGFRB or FGFR1.

Prognosis: The prognosis of aCML is poor with a median survival of less than two years. Approximately one-third progress to AML, while the remainder succumb to BM failure.

Juvenile Myelomonocytic Leukemia

Juvenile myelomonocytic leukemia (JMML) is a clonal hematologic stem cell neoplasm of children characterized predominantly by proliferation of granulocytic and monocytic cells. The incidence is slightly more than one case per million children (<14 years old) with a male-to-female ratio of 2:1. Approximately 75% of cases occur in children younger than three years of age. About 10% of cases occur in children with neurofibromatosis type 1 (NF-1).

Clinical findings: Common clinical features include malaise, fever, bleeding and infections such as bronchitis and tonsillitis. Nearly all patients have hepatosplenomegaly. Maculopapular skin rashes due to infiltration by leukemic cells occur in nearly half of patients. Café-au-lait spots are common in the patients with NF-1.

Peripheral blood findings: The PB demonstrates elevation of the WBC (median range 25-35 x 10⁹/L) due to neutrophilia and monocytosis. There is a granulocytic left-shift. Blasts and promonocytes usually comprise <5% of the PB differential and, by definition, must be <20%. Nucleated RBCs are common.

Table 5.4. WHO diagnostic criteria for JMML

1. PB monocytosis >1 x 10⁹/L
2. Blasts and promonocytes <20% of WBCs in PB and BM
3. BCR/ABL absent
4. Plus two or more of the following:
   - Hgb F increased for age
   - Immature granulocytes in PB
   - WBCs >10 x 10⁹/L
   - Clonal chromosomal abnormality
   - GM-CSF hypersensitivity of myeloid progenitors in vitro

Adapted with permission from: Swerdlow SH, Campo E, Harris NL et al. World Health Organization Classification of Tumours of Haematopoietic and Lymphoid Tissues. Lyon: IARC, 2008:82, Table 4.03.

Anemia is variable, while thrombocytopenia is common. Eosinophilia and basophilia are sometimes seen. Dysplastic changes may be present but are not prominent.

Bone marrow findings: The BM is hypercellular due to granulocytic proliferation with moderately increased monocytic cells (5-30%). By definition, BM blasts are <20% and Auer rods are absent. Promonocytes are considered as blast equivalents. The megakaryocytes are often decreased. Dysplasia is minimal or absent.

Laboratory findings: The laboratory abnormalities of JMML are evident in the WHO criteria depicted in Table 5.4. The formation of spontaneous granulocyte-macrophage colonies in vitro and the hypersensitivity of these colonies to granulocyte macrophage colony stimulating factor (GM-CSF) are important studies for confirmation of the diagnosis. Elevated fetal hemoglobin occurs in two-thirds of patients and is an important clue to the diagnosis in the early course as well as an important prognostic feature. Polycolonal hypergammaglobulinemia and elevated serum lysozyme are seen in a majority of cases.

Genetics: By definition, BCR/ABL is absent in JMML. Abnormalities on cytogenetic studies occur in about 45% of cases, including monosomy 7. RAS mutations are seen in approximately 20% of patients, which can be detected by molecular studies and may be associated with a worse prognosis. Children carrying the diagnosis of NF-1 have a 200-to-500-fold increased risk for developing a myeloid neoplasm, usually JMML.

Infectious mimics of JMML: There are a number of infectious diseases with laboratory and clinical features that closely resemble JMML. When clonality cannot be demonstrated by either cytogenetic or molecular studies, then it is important to rule out Epstein-Barr virus, cytomegalovirus, human herpesvirus 6, histoplasma, mycobacteria and toxoplasma.

Prognosis: The clinical course of JMML is variable, but most patients eventually succumb to the disease, usually due to organ failure caused by leukemic infiltration. Approximately 30% of patients have an early progressive disease course with rapid death. About 10% to 20% of patients progress to AML. The remaining patients have an indolent, slower-progressing disease course regardless of therapy. Favorable and unfavorable prognostic factors are given in Table 5.5. Reported median survival time ranges from five months to greater that four years. Bone marrow transplantation is the only therapy that offers a potential cure.

Table 5.5. Prognostic factors in JMML

Favorable	Unfavorable
<1 year of age	>2 years of age
	Platelets <33 x 10⁹/L
	Hgb F level >15% at time of diagnosis
	Mutation of RAS

Myelodysplastic/Myeloproliferative Neoplasms, Unclassifiable

Myelodysplastic/myeloproliferative neoplasms, unclassifiable (MDS/MPN, U) are disorders in which there is at least one myeloid cell line demonstrating ineffective hematopoiesis and at least one myeloid cell line with effective hyperproliferation, that does not fit the criteria for any of the previously described MDS/MPNs. The morphology and clinical features vary depending on which cell lines are involved and in what way. The WHO diagnostic criteria for MDS/MPN, U are given in Table 5.6. The prognosis is variable.

An important element of the WHO diagnostic criteria is that there is no prior diagnosis of MPN or MDS. In particular, it should be kept in mind that a previously unrecognized MPN may be undergoing progression or transformation. Intimate with this concept is the requirement that BCR/ABL is absent.

MDS/MPN, U-RARS associated with marked thrombocytosis: Approximately 15% of patients that have the clinical and morphologic features of the MDS, refractory anemia with ring sideroblasts (RARS), have a platelet count >500 x 10⁹/L. These cases may be placed in the MDS/MPN, U category; however, it has been proposed that these cases be placed in a provisional entity, MDS/MPN,

Table 5.6. WHO diagnostic criteria for MDS/MPD,U

- The case has clinical, laboratory, and morphologic features of one of the recognized categories of MDS, with <20% blasts in the PB and BM.

AND

- Has prominent myeloproliferative features such as plt >450 x 10⁹/L with megakaryocytic proliferation, or WBC >13 x 10⁹/L with or without splenomegaly

AND

- Has no prior history of CMPD or MDS, no recent cytotoxic or growth factor therapy that can explain the findings, no BCR/ABL, no rearrangement of PDGFRA, PDGFRB or FGFR1 and no del(5q), t(3;3)(q21;q25), or inv(3)(q21;q26)

OR

- The patient has de novo disease with mixed MDS and MPN features that cannot be assigned to any other category of MDS, MPN, or MDS/MPN

Adapted with permission from: Swerdlow SH, Campo E, Harris NL et al. World Health Organization Classification of Tumours of Haematopoietic and Lymphoid Tissues. Lyon: IARC, 2008:85, Table 4.04.

U-RARS associated with marked thrombocytosis. The definition of this entity requires the absence of the del(5q) and 3q21q26 chromosomal abnormalities. Those with del(5q) should be placed in the MDS category of 5q- syndrome, and those with 3q21q26 should be diagnosed as either MDS or AML, according to whether the blast count is less or more than 20%.

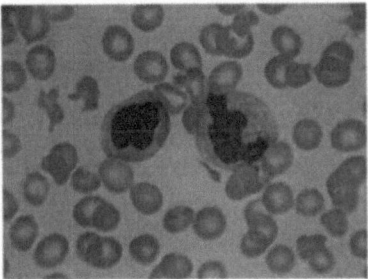

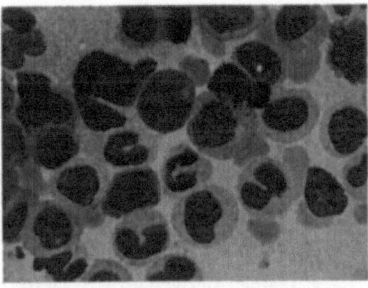

Figure 5.1. CMML, peripheral blood showing abnormal monocytes (500x).

Figure 5.2. CMML, BM aspirate with abnormal and immature monocytic cells in a background of maturing granulocytes (500x).

CHAPTER 6

Myeloid Neoplasms with Eosinophilia with Abnormalities of PDGFRA, PDGFRB or FGFR1

Introduction

This newly defined group of neoplasms, not included in the 2001 edition of the WHO bluebook, consists of three clonal entities associated with eosinophilia. In the previous WHO classification, these disorders were usually placed under chronic eosinophilic leukemia (CEL), mast cell disease (MCD) or atypical chronic myeloid leukemia (aCML), depending on the predominant features. Modern molecular techniques have identified these subsets of patients with clonal eosinophilia that are associated with mutations of the following tyrosine kinase genes:

- Platelet-derived growth factor receptor A (PDGFRA)
- Platelet-derived growth factor receptor B (PDGFRB)
- Fibroblast growth factor receptor 1 (FGFR1)

Because these chromosomal abnormalities have clinical and therapeutic significance, they are now recognized as distinct entities.

PDGFRA Mutation

The PDGFRA abnormality involves a juxtaposition of the PDGFRA and FIP1L1 genes (FIP1L1-PDGFRA) and is present in approximately 50% of patients with systemic mastocytosis (SM) that are associated with prominent eosinophilia (SM-eos), which account for approximately 20% of all cases of SM in adults. These patients exhibit a "myeloproliferative" clinical picture with organomegaly, elevated serum B12 levels and bone marrow (BM) hypercellularity with fibrosis. Traditionally, some of these cases were diagnosed as CEL; however, mast cell infiltrates, if diligently searched for, can be found in most. Rarely, PDGFRA is activated through the t(4;22)(q12;q11) abnormality. Patients with FIP1L1-PDGFRA positive myeloid neoplasms respond to imatinib (Gleevac™) therapy.

PDGFRB

The PDGFRB gene is activated through several translocations (Table 6.1). These patients have a "myeloproliferative" picture with marked eosinophilia and sometimes with monoytosis. Some are diagnosed as CEL or chronic myelomonocytic leukemia (CMML). PDGFRB neoplasms respond well to (Gleevac™) therapy.

Bone Marrow: A Practical Manual, by Daniel A. Cherry and Tomislav M. Jelic.
©2011 Landes Bioscience.

Table 6.1. Translocations that activate PDGFRB and FGFR1

PDGFRB	FGFR1
t(5;12)(q33;p13)	t(8;13)(p11;q12)
t(5;10)(q33;q21)	t(8;9)(p11;q33)
t(5;7)(q33;q11.2)	t(6;8)(q27;p11)
t(5;14)(q33;q13)	t(8;22)(p11;q22)
t(5;17)(q33;p11)	
t(1;5)(q23;q33)	

FGFR1

FGFR1is located on chromosome 8p11 and is activated through several translocations (Table 6.1). These patients have a "myeloproliferative" picture with prominent eosinophilia, lymphadenopathy and an increased incidence of T-cell lymphoma with progression to acute leukemia. Other variants involving t(8;17) (p11;q25), t(8;11)(p11;p15), t(8;12)(p11;q15) and ins(12;8)(p11;p11) have been associated with T-cell lymphoma and BM eosinophilia.

BM Evaluation

BM sampling in patients with PDGFRA-, PDGFRB-, and FGFR1-associated neoplasms is usually prompted by eosinophilia in which reactive etiologies have been ruled out (see discussion of CEL in chapter on MPNs). Because of the treatment implications of the presence of PDGFRA and PDGFRB, awareness of the existence of these disorders is critical. The translocations associated with PDGFRB and FGFR1 can be detected by standard metaphase cytogenetics. Identification of the FIP1L1-PDGFRA abnormality requires FISH. Immunohistochemical stains for tryptase and CD25, the latter of which is positive in neoplastic but not in non-neoplastic mast cells, can be useful for the identification of subtle mast cell infiltrates in the BM biopsy.

CHAPTER 7

Acute Myeloid Leukemia

Introduction

Acute myeloid leukemias (AMLs) are clonal malignancies of hematologic cells of granulocytic, monocytic, erythroid and megakaryocytic lineage. The morphologic findings in the peripheral blood (PB) consist of variable cytopenias with variable numbers of blasts, and in the bone marrow (BM) of hypercellularity with, by definition, increased blasts. The clinical symptoms are usually related to cytopenia(s), cell dysfunction or tumor burden. Examples are:

- Fatigue, chest pain, shortness of breath and other symptoms secondary to anemia
- Bleeding due to low platelets
- Infection because of dysfunction of the white blood cells (WBCs) or neutropenia
- Thrombosis or hyperviscosity due to a large number of blasts in the blood
- Bone pain caused by the massive expansion of cells in the BM
- Disseminated intravascular coagulation (DIC) related to cytokine release by the malignant cells

Basic steps in the work-up of AML: When a hematologic malignancy is suspected, the following series of steps should be taken to establish a diagnosis of AML.

1. Identify abnormal cells in the BM or PB as blasts
2. Characterize the blast lineage
3. Determine whether or not the criteria for AML have been met
4. Classify the AML

The sequence and time frame in which information becomes available to the pathologist can seem chaotic. The diagnosis of acute leukemia can be made within a few hours based on increased blasts in the BM aspirate or PB. Determination of myeloid lineage, in the absence of Auer rods, requires flow cytometry (FCM) or cytochemistry, and the "final" diagnosis may not be known until genotyping (a few days to a week) is complete. The pathologist's job is to manage these events and to interpret, or refuse to over-interpret, the results as they become available. In many instances the pathologist can give the clinician enough information to provide the patient with a provisional explanation of their disease and to begin treatment quite early in this process.

Blast morphology is variable. They may be slightly larger or many times larger than normal lymphocytes with nuclear chromatin that is fine or dense. The nuclei can show mild to extreme irregularities of shape and zero to several nucleoli. The cytoplasm is basophilic or deeply basophilic, scant to abundant and sometimes contains granules or one to many Auer rods. Blasts of the same lineage in the same

Bone Marrow: A Practical Manual, by Daniel A. Cherry and Tomislav M. Jelic.
©2011 Landes Bioscience.

Table 7.1. Immunophenotypic markers by FCM in AMLs

Cell Type	Markers
Hematologic precursors	CD34, CD45, HLA-DR, TdT
Myeloid	CD13, CD33, CD15, MPO, CD16, CD117
Monocytic	CD14, CD11b, CD11c, CD4, CD64, CD36
Erythroid	Glycophorin A, Hemoglobin A
Megakaryocytic	CD41, CD61
B-lymphocytic	CD19, CD20, CD22, CD79a
T-lymphocytic	CD2, CD3, CD7, CD5

MPO: myeloperoxidase

specimen may differ in appearance. Specific characteristics of blasts of different lineage are described with the corresponding AMLs.

Myeloblast versus lymphoblast: In the absence of Auer rods a myeloblast cannot be reliably differentiated from a lymphoblast or a high-grade lymphoma cell by morphologic criteria on a Wright-Giemsa (WG) stained smear. Morphology should be practiced but not completely trusted. The presence of cytoplasmic granules does not assure myeloid lineage. Because the treatment and prognosis of AML and acute lymphoblastic leukemia (ALL) differ, it is not wise to tell the clinician the lineage of the blasts before it has been proven by FCM or cytochemistry.

Special clinical implications of acute promyelocytic leukemia (APL): No discussion of the initial evaluation of acute leukemia is adequate without the warning that APL is a medical emergency due to associated coagulopathy, and any suspicion of this diagnosis should immediately trigger an evaluation for DIC and fluorescence in situ hybridization (FISH) studies to confirm the diagnosis by identification of the t(15;17) chromosomal abnormality. A more detailed description of APL is given below.

Blast lineage by flow cytometry (FCM): FCM is currently the preferred technique for characterization of blast lineage and should be done on all cases in which acute leukemia is suspected. Markers for several cell types important in the evaluation for AML are presented in Table 7.1. A more in-depth discussion of FCM is presented in Chapter 2.

Cytochemistry (Fig. 7.1): Cytochemistry is currently used less frequently than in the past because it yields less information and because interpretation may be challenging. For example, differentiating a myelocyte from a blast can be problematic when there is strong chloracetate esterase (CAE) positivity that obscures the morphology. The usefulness of cytochemistry, however, should not be taken for granted. It is a rapid (about one hour), inexpensive and reliable way to demonstrate myeloid lineage. Table 7.2 lists commonly used cytochemical stains and expected reactivity. A popular technique is the dual esterase stain, which combines CAE and nonspecific esterase (NSE) on the same slide.

Blast equivalents: In some AMLs, cells other than myeloblasts are considered as blast equivalents for diagnostic purposes. These include promyelocytes in APL, promonocytes and monoblasts in acute monocytic/monoblastic leukemia, and

Table 7.2. Cytochemical stains used in evaluating AMLs

Cytochemical Stain	Findings
Myeloperoxidase (MPO)	Positive in granulocytic cells; may be negative in abnormal granulocytes
Sudan black B (SBB)	Positive in granulocytic cells; may be weakly positive in monocytic cells and rarely in ALL
Chloracetate esterase (CAE) (naphthol ASD chloracetate esterase)	Positive in granulocytes; positive in granular ALL; positive in eosinophils in acute myelomonocytic leukemia; positive in mast cells
Nonspecific esterase (NSE) (alpha-naphthyl acetate esterase and alpha-naphthyl butyrate esterase)	Positive in monocytic cells; may be weakly positive in granulocytes
Periodic acid-Sciff (PAS)	Large globule positivity in malignant erythroblasts and lymphoblasts; diffuse cytoplasmic positivity in more mature neoplastic erythroid cells

megakaryoblasts in acute megakaryoblastic leukemia. Furthermore, as will be seen, the blast percentage criteria are a bit different when diagnosing acute myelomonocytic leukemia and acute leukemias with erythroid lineage.

Low blast-count AML: Cases with increased blasts, but less than the 20% normally required, are still diagnosed as AML when the t(8;21) or t(16;16) chromosomal abnormalities are present. These would, otherwise, be classified as a myelodysplastic syndrome (MDS).

World Health Organization (WHO) classification: The gold standard for the classification of AML is the WHO system, which has seven categories:

1. AML with recurrent chromosomal abnormalities
2. AML with MDS-related changes
3. Therapy-related myeloid neoplasms
4. AML, NOS (not otherwise specified)
5. Myeloid sarcoma
6. Myeloid proliferations related to Down syndrome
7. Blastic plasmacytoid dendritic cell neoplasm

Terminology from the older French American British (FAB) classification is still frequently heard, mostly because the FAB names are succinct, for example, M2 for acute myeloid leukemia with maturation. The most conspicuous change from the FAB to the WHO system is the lowering of the required blast count from 30% to 20%. FAB was based primarily on blast lineage as determined with morphology and cytochemistry, while the WHO attempts to take into account all of the immunophenotypic, genetic and clinical features which collectively distinguish a disease as a unique entity. FAB terminology will be correlated with the WHO classes where applicable.

Genetic studies: Chromosomal analysis (cytogenetics or CG) should be performed on all cases of suspected acute leukemia. In addition to those diseases that are defined by their chromosomal abnormalities, the presence or absence of other chromosomal defects are important prognostic findings. FISH and PCR can be performed more rapidly than cytogenetics and are more sensitive for some defects. A disadvantage is that these methods use probes that are unique for the genes being sought and, therefore, one must know what he is looking for ahead of time. Because FISH and PCR are expensive and often redundant with cytogenetics, their specific clinical relevance should be carefully considered on a case by case basis. A more detailed discussion of these techniques is presented in the Chapter 2.

Acute Myeloid Leukemia, Not Otherwise Specified

This group of leukemias is presented first because they are defined using the fundamental properties that serve as the basis upon which the current more sophisticated knowledge of AMLs is built. Historically, most of the AMLs in the other groups were given one of these diagnoses prior to our modern understanding of immunophenotyping and genetics. This section should be referred to for fundamental morphologic, immunohistochemical (IHC), cytochemical and immunophenotypic descriptions of the major cell types.

AML, Minimally Differentiated, FAB M0

In this disease, the blasts cannot be classified as myeloid based on morphology or light microscopic cytochemistry. Less than 3% of the blasts are positive for MPO, CAE, SBB or NSE. Differentiation from ALL can be difficult in many instances because the blasts are small and resemble lymphoblasts. This is further complicated by the fact that the blasts are positive for TdT in up to one-half of cases. Lineage is established by the demonstration of myeloid antigens using FCM. This class of AML is probably underdiagnosed because cytochemistry is frequently not attempted and, therefore, not found to be ambiguous. Important features are summarized in Table 7.3.

AML without Maturation FAB M1 (Figs. 7.1, 7.2)

Blasts comprise ≥90% of the non-erythroid cells in the BM and ≥3% of the blasts are positive for MPO, CAE or SBB. In modern practice, cytochemistry is not always done and the blasts are characterized by demonstrating at least two myeloid markers by FCM. They are usually negative for lymphoid and monocytic markers. The blasts are commonly well differentiated, containing a large nucleus with finely dispersed chromatin and a prominent nucleolus. The nucleus to cytoplasmic ratio (N:C) is high, and there are variable numbers of cytoplasmic granules or Auer rods. Sometimes the blasts are poorly differentiated and can be mistaken for lymphoblasts. Important features are summarized in Table 7.4.

AML with Maturation FAB M2 (Fig. 7.3)

Acute myeloid leukemia with maturation contains 20% to 89% blasts and <20% monocytic cells in the BM. The blasts are frequently well differentiated with cytoplasmic granules and often Auer rods. This is the category into which most cases of AML with the recurrent t(8;21) abnormality fall morphologically, thus,

Table 7.3. Important features of AML, minimally differentiated

FAB	M0
Lineage	Granulocytic
Definition	No evidence of myeloid differentiation by light microscopy including cytochemistry
Frequency	<5% of AMLs
Cytochemistry	Negative for MPO, SBB, CAE, NSE
FCM	Positive for CD13, CD33, CD117
	MPO often negative
	Usually positive for CD34, CD38, HLA-DR
	Usually negative for monocytic antigens
	TdT positive in 1/2 of cases
	May express nonspecific lymphoid antigens (CD7, CD2, CD19)
IHC	Usually negative for MPO
	May be positive for CD34
	TdT positive in 1/2 of cases
Genetics	Nonspecific
	Most commonly complex (>3 recurring) abnormalities
Special features	Blasts sometimes resemble lymphoblasts
Prognosis	Poor

Table 7.4. Important features of AML without maturation

FAB	M1
Lineage	Granulocytic
Definition	≥90% of non-erythroid cells in BM are blasts
Frequency	10% of AMLs
Cytochemistry	≥3% of blasts positive for MPO, SBB, CAE
FCM	Positive for one or more CD13, CD33, CD117, CD34, HLA-DR
	Usually positive for MPO
IHC	MPO positive
Genetics	Nonspecific
Special features	Rarely, blasts resemble lymphoblasts but are TdT negative
Prognosis	Poor

Table 7.5. Important features of AML with maturation

FAB	M2
Lineage	Granulocytic
Definition	20%-89% blasts and <20% monocytic cells in BM
Frequency	10% of AMLs
Cytochemistry	Positive for MPO, SBB, CAE
FCM	One or more of CD13, CD33, CD15, CD65, CD11b expressed
	May express CD117, CD34, HLA-DR
	CD7 expressed in 25% but expression of other lymphoid or monocytic markers is rare
IHC	MPO positive
Genetics	No associated recurrent abnormalities
Special features	Variable dysplasia may be present including hyposegmented neutrophils (pseudo-Pelger-Huet cells)
	Eosinophils frequently elevated, but there is no abnormality of chromosome 16
Prognosis	Intermediate; less favorable than with t(8;21)

serving as an example of why a final diagnosis cannot be given until genetic studies are complete. Important features are summarized in Table 7.5.

Acute Myelomonocytic Leukemia FAB M4

AML accompanied by increased monocytic cells. There must be ≥20% blasts and ≥20% monocytic cells, including promonocytes and monoblasts. Promonocytes are large cells with immature monocytic appearing nuclei with delicate nuclear folding and abundant blue-gray cytoplasm that may be vacuolated and may contain fine granules. Monoblasts are slightly larger with a slightly greater N:C. The nucleus may be round or folded, and the cytoplasm is abundant, sometimes with fine granulation or vacuolization (see Fig. 7.5). There may be morphologic, cytochemical and immunophenotypic overlap between myeloblasts, promonocytes and monoblasts. This is a disease in which cytochemistry may be quite helpful for delineating the myeloid from monocytic lineage of the cells. Dual esterase staining demonstrates positivity in the monocytic cells for NSE and in the myeloid cells for CAE. Some cells may show dual positivity. Flow cytometry, likewise, can show clearly separate populations of myeloid and monocytic cells or an overlapping immunophenotypic spectrum of myeloid and monocytic features. This can make accurate cell counting difficult. Cases with abnormalities of chromosome 16 or 11q23 are excluded from this group. Important features are summarized in Table 7.6.

Table 7.6. Important features of acute myelomonocytic leukemia

FAB	M4
Lineage	Granulocytic and monocytic
Definition	≥20% blasts and ≥20% monocytic cells
Frequency	5%-10% of AMLs
Cytochemistry	Myeloid component positive for MPO, SBB, CAE
	Monocytic component positive for NSE
	Some cells may be positive for both myeloid and monocytic markers
FCM	Myeloid positive for CD13, CD33, CD65, CD16
	Monocytic positive for CD14, CD4, CD11b, CD11c, CD64, CD36, CD68, CD163
	Co-expression of CD15 and CD64 is characteristic
	Blasts often positive for CD34 and/or CD117
	Usually HLA-DR positive
	30% positive for CD7
IHC	Myeloid cells positive for MPO
	Monocytic cells positive for CD68 and CD163
Genetics	Nonspecific
	+8 present in most cases
	Must be negative for chromosome 16 and 11q23 abnormalities
	t(8;16)(p11.2;p13.3) associated with hemophagocytosis
Special features	May be morphologic, cytochemical and immunophenotypic overlap of myeloid and monocytic features
Prognosis	Intermediate
	Less favorable than cases with chromosome 16 abnormalities

Acute Monoblastic Leukemia and Acute Monocytic Leukemia (Fig. 7.5)

These leukemias are characterized by the presence of ≥80% monocytic cells including monocytes, promonocytes and monoblasts. If ≥80% of the monocytic cells are monoblasts, then the designation *acute monoblastic leukemia* is used. Can be associated with gingival, cutaneous and central nervous system (CNS) involvement. Only cases lacking 11q23 abnormalities are included in this group. Cases with t(8;16)(p11.2;p13.3) are associated with hemophagocytosis. Important features are summarized in Table 7.7.

Table 7.7. Important features of acute monoblastic and acute monocytic leukemia

FAB	M5
Lineage	Monocytic
Definition	≥80% monocytic cells
	If ≥80% of the monocytic cells are monoblasts, then acute monoblastic leukemia
Frequency	Acute monocytic leukemia <5% of AMLs
	Acute monoblastic leukemia <5% of AMLs
Cytochemistry	Generally positive for NSE, but up to 10%-20% of cases may be negative
FCM	Always positive for some monocytic markers, but less frequently express mature monocytic markers such as CD14 and more commonly express less mature markers such as CD36, CD64, CD4, CD11b, CD11c, VD68, CD36
	Variably express myeloid markers CD13, CD33, CD15, CD65
	May be positive for MPO in monocytic and less often in monoblastic
	CD34 frequently negative, CD117 positive more common and CD33 often very bright
IHC	Lysozyme positive
	MPO usually negative
	May be CD68 and/or CD163 positive
Genetics	t(8;16)(p11;p13) associated with hemophagocytosis and coagulopathy
	Must lack 11q23 abnormality
Special features	Associated with gingival, cutaneous and CNS involvement
Prognosis	Poor

Acute Erythroid Leukemias (Fig. 7.6)

Two diseases are placed in this category:

- **Erythroleukemia** demonstrates ≥50% of the marrow cells to be erythroid and ≥20% of the non-erythroid cells to be myeloblasts.
- In **pure erythroid leukemia** >80% of the immature neoplastic cells are erythroid.

The erythroid cells in erythroleukemia show marked dysplasia, sometimes with ring sideroblasts. In fact, some of these cases evolve from MDS. The erythroid cells may contain cytoplasmic vacuoles that are positive for PAS.

The myeloblastic component is morphologically, cytochemically and immunophenotypically similar to that described for other AMLs. In pure erythroid leukemia the blasts can have the appearance of large immature erythroid cells or they can be small and similar in appearance to lymphoblasts, but are negative for TdT. This disease can resemble megaloblastic anemia, however, in megaloblastic anemia there are also changes in the granulocytic and megakaryocytic cells, and the blasts rarely exceed 5%. If the diagnosis remains in doubt, then the patient should be given a course of B12 and folate therapy before a diagnosis of leukemia is made. Important features of acute erythroid leukemias are summarized in Table 7.8.

Acute Megakaryoblastic Leukemia (Figs. 7.8, 7.9)

This leukemia demonstrates ≥20% myeloblasts and ≥50% of the blasts to be of megakaryocytic lineage and can be associated with many unusual clinical features. In children there may be a prominent abdominal mass and an association with the t(1;22) abnormality, which is a separate entity in the 2008 revised WHO classification (see below). Some patients show lytic bone lesions. In young males there is an association with mediastinal germ cell tumors. Splenomegaly is uncommon. Thrombocytopenia is common and thrombocytosis may occur rarely. Megakaryoblasts are highly variable in appearance and are frequently recognizable without special studies. They are sometimes quite large with mildly irregular nuclei and cytoplasmic pseudopods, or they may be small and lymphoid-appearing. They are negative for TdT. The PB can contain dysplastic platelets and micromegakaryocytes as well as circulating megakaryoblasts and fragments. The appearance of the BM biopsy may be difficult to distinguish from metastatic carcinoma. Immunohistochemistry for Factor VIII (FVIII), CD41, CD61 and cytokeratins is helpful in such instances. Important features are summarized in Table 7.9.

Acute Basophilic Leukemia

This is a very rare leukemia composed primarily of basophils. Clinical features can include symptoms from hyperhistaminemia such as:

- Urticaria
- Skin lesions
- Organomegaly
- Lytic bone lesions

The malignant cells are immature basophils possessing oval or bilobed nuclei with nucleoli and coarse basophilic granules in the cytoplasm. Cases with mast cell differentiation have a characteristic appearance in the BM sections consisting of paratrabecular groups of elongated cells with oval nuclei, often cuffed by eosinophils. The malignant basophils are positive on toluidine blue staining. Important features are summarized in Table 7.10.

Acute Panmyelosis with Myelofibrosis

This is a very rare de novo acute leukemia which is accompanied by BM fibrosis. There is always pancytopenia. In contrast to primary myelofibrosis (PM) (see Chapter 4), splenomegaly is minimal or absent. The BM aspirate is frequently

Table 7.8. Important features of acute erythroid leukemias

FAB	M6
Lineage	Erythroid and myeloid in erythroleukemia
	Erythroid in pure erythroid leukemia
Definition	In erythroleukemia, ≥50% erythroid cells and ≥20% of the non-erythroid cells are blasts
	In pure erythroid leukemia, >80% erythroid cells with no significant myeloblastic component
Frequency	Erythroleukemia <5% of AMLs
	Pure erythroid leukemia is extremely rare; sometimes occurs as a childhood leukemia
Cytochemistry	Erythroblasts show cytoplasmic PAS positivity in large granules and globules and are MPO negative
	More mature neoplastic erythroid cells show diffuse cytoplasmic PAS positivity
	Myeloblasts in erythroleukemia are positive for MPO, SBB, CAE
FCM	More differentiated erythroid cells positive for CD71, glycophorin A (Gly A) and hemoglobin A (Hgb A), and negative for MPO
	More immature erythroid cells are negative for Gly A and Hgb A and are positive for carbonic anhydrase, Gero antibody against the Gerbich blood group, and CD36
	Negative for megakaryocytic markers CD41 and CD61
	Erythroid cells CD45 negative
	Myeloblastic component of erythroleukemia demonstrates typical myeloid antigens
IHC	Erythroid cells positive for Gly A , Hgb A
	Myeloblasts positive for MPO
Genetics	Nonspecific
	Complex (>3 abnormalities) karyotype common, most commonly involving chromosomes 5 and 7
Special features	May occur de novo or evolve from MDS or chronic myeloid leukemia
	May have ring sideroblasts
	Blasts in pure erythroid leukemia, rarely, resemble lymphoblasts but are TdT negative
Prognosis	Poor

7

Table 7.9. Important features of acute megakaryoblastic leukemia

FAB	M7
Lineage	Megakaryocytic
Definition	≥20% blasts and ≥50% of blasts are of megakaryocytic lineage
Frequency	<5% of AMLs
Cytochemistry	May show positivity for PAS in a fashion similar to that of erythroblasts
	May show punctate NSE positivity
	Negative for MPO, SBB
FCM	Positive for CD41 and/or CD61
	Positive for CD36
	May be positive for CD13, CD33
	Often negative for CD34, CD45, HLA-DR
IHC	Positive for FVIII
	CD61 positivity dependent on fixation and decalcification techniques
	TdT negative
Genetics	Some cases show inv(3)
	i(12p) associated with mediastinal germ cell tumors in young males
	t(1;22) abnormality is absent
Special features	May show lytic bone lesions or be associated with mediastinal germ cell tumor
	May morphologically mimic metastatic carcinoma in BM biopsy
Prognosis	Poor

unsuccessful (dry tap). The BM biopsy is densely fibrotic and hypercellular with panmyelosis including groups of blasts (median blast count 22.5%) and prominent megakaryocytes. Further distinctions from PM include the prominence of immature cells and hypolobated megakaryocytes, as opposed to the hyperlobation seen in PM. The distinction from acute megakaryoblastic leukemia and AML with myelodysplasia-related changes is rather arbitrary and probably clinically irrelevant. FCM, cytogenetics and cytochemistry are usually not possible due to a dry tap. Flow cytometry of morcellated, non-decalcified BM biopsy fragments may be attempted primarily for the purpose of obtaining an accurate blast count. Important features are summarized in Table 7.11.

Table 7.10. Important features of acute basophilic leukemia

FAB	None
Lineage	Basophilic
Definition	Acute leukemia in which the primary differentiation is to basophils
Frequency	<1% of AMLs
Cytochemistry	Metachromatic positivity for toluidine blue
	Some cases are PAS positive in block pattern
	Blasts often negative for MPO, SBB, NSE
FCM	Blasts positive for CD13, CD33
	Blasts usually positive for CD9, CD123, CD203c, CD11b and maybe for CD34 and HLA-DR (negative in normal basophils)
	Negative for CD117 (differentiates from mast cells)
	Some cases positive for TdT
IHC	Cases with mast cell differentiation positive for CD117
Genetics	Nonspecific
	12p and t(6;9), which are associated with basophilia in some AMLs, are not present
	BCR/ABL positive cases are excluded
Special features	May have cutaneous lesions, organomegaly, lytic bone lesions, or symptoms of hyperhistaminemia
Prognosis	Unknown because of rarity

Acute Myeloid Leukemia with Recurrent Genetic Abnormalities

This group of leukemias was created in recognition of the importance of certain recurrent chromosomal abnormalities that are associated with specific patterns of disease with primarily prognostic implications. It is notable that genetic abnormalities are being characterized at a rate so rapid as to make the 2008 revision of the WHO classification obsolete before the ink has dried on its pages. Modern techniques, such as genetic profiling using microarray analysis, have the potential to make morphology an interesting observation accompanying classification based primarily on genetics.

All of the diseases in the AML with recurrent genetic abnormalities group have correlates in the FAB system, and all but one, APL, have correlates in the WHO AML, NOS group. The previous discussion of AML, NOS, should be referred to for details regarding the general characteristics of the major cell lines.

With the exception of APL, the chromosomal abnormalities are mostly of prognostic significance and do not alter initial treatment. Therefore, to avoid a delay in treatment, a provisional diagnosis using the AML, NOS terminology should be made with a comment that genetic studies are pending.

Table 7.11. Important features of acute panmyelosis with myelofibrosis

FAB	None
Lineage	Granulocytic with some cases showing some megakaryocytic and/or erythroid differentiation
Definition	Acute leukemia accompanied by BM fibrosis
Frequency	Very rare
Cytochemistry	Usually not possible because of dry tap
	Sometimes megakaryocytes are prominent and are positive for PAS
FCM	Usually not possible because of dry tap
	Can be attempted on non-decalcified, morcellated BM biopsy material
	May be positive for myeloid, erythroid, and/or megakaryocytic antigens in proportion to the presence of these cells
IHC	Positive for MPO, CD41, CD61, FVIII, Gly A, Hgb A relative to proportions of myeloid, megakaryocytic and erythroid cells present
Genetics	Usually not possible due to dry tap
	The presence of complex (>3 recurring) abnormalities of chromosomes 5 and/or 7 require classification as AML with myelodysplasia-related changes
Special features	Acute onset with constitutional symptoms, bone pain and fever.
	Heterogeneity of blasts as opposed to a predominant lineage
Prognosis	Poor

Cytogenetically normal AMLs with genetic mutations: These entities, newly added in the 2008 WHO revision of the AML with recurrent genetic abnormalities group, have a normal karyotype by standard metaphase cytogenetics (AML-NK) but contain genetic mutations by molecular testing. They comprise a large proportion of AMLs and are the subject of ongoing research that is uncovering important differences in treatment response. The implications of most of these differences have not yet trickled down to the community hospital level, but expect them to do so very soon. Included are:

- Mutated/cytoplasmic NPM
- Mutation of CEBPA
- Translocations of MLL

Another recently discovered cryptic genetic abnormality in AML is mutated FLT3, which does not define a new entity in the WHO revision but is discussed below because of its prognostic importance. Interestingly, stratification of cases of AML-NK using mutations of NPM CEBPA, and FLT3 may have prognostic relevance independent of whether the AML is therapy related or has evolved from MDS (see descriptions of these below).

AML with t(8;21)(q22;q22); RUNX1-RUNX1T1

This leukemia is most commonly morphologically equivalent to AML with maturation (M2), but demonstrates the t(8;21)(q22;q22), which is the most common chromosomal abnormality seen in AML by cytogenetics (5% to 12% of cases). Rare cases demonstrate M1 or M4/M5 morphology. Auer rods are common in the blasts and may also be seen in mature neutrophils. Pseudo-Pelger-Huet cells and neutrophils with large granules (pseudo-Chediak-Higashi cells) are sometimes present in the PB. Eosinophilia can be seen, but not to the extent as in acute myelomonocytic leukemia with eosinophilia (see below). Other cases may have increased basophils and/or mast cells. Some patients present with granulocytic sarcoma (malignant extramedullary myeloid tumor). The blast count is elevated but may not be 20%; however, these cases should still be classified as AML rather than MDS. By FCM these cases show myeloid antigens such as CD13, CD33 and MPO and they frequently co-express the B-lymphocytic antigens CD19 and PAX5. The prognosis is superior to that of equivalent leukemias without t(8;21). CD56 is sometimes expressed and is a poor prognostic finding.

AML with inv(16)(p13.1;q22) or t(16;16)(p13.1;q22); CBFB/MYH11 (Figs. 7.10, 7.11)

This entity, referred to as acute myelomonocytic leukemia with abnormal marrow eosinophils (AMMLeo), is morphologically equivalent to AMML (M4) with the addition of increased abnormal eosinophils. These genetic defects occur in about 5% to 8% of AMLs. The abnormal eosinophils demonstrate the full range of maturation, with large purple granules seen mostly in the eosinophilic promyelocyte and eosinophilic myelocyte stages. The immature eosinophils may demonstrate hypolobation. Chloracetate esterase is weakly positive in the eosinophils in contrast to normal eosinophils that are characteristically negative. Rare cases lack the eosinophilic component. Mature neutrophils are decreased in the BM. Some cases have slightly less than 20% blasts but should still be diagnosed as AML. FISH and PCR are more sensitive for detecting these chromosome 16 abnormalities and should be done in all cases in which this disease is suspected based on morphology because of the therapeutic significance. Some patients present with myeloid sarcoma. The prognosis is superior to equivalent leukemias lacking the chromosome 16 abnormalities and is made better by treatment with high-dose cytarabine given during the consolidation phase of therapy.

Acute Promyelocytic Leukemia with t(15;17)(q22;q12); PML-RARA (Fig. 7.12)

This disease is designated as M3 in the FAB system and is characterized by the presence of abnormal promyelocytes with intense cytoplasmic granulation and sometimes multiple Auer rods (called faggot cells because of resemblance to stacks of fire wood). The microgranular variant is composed of smaller cells with decreased cytoplasmic granulation and folded, sometimes bilobed nuclei that are easily confused with monocytic cells (Fig. 7.13). Acute promyelocytic leukemia accounts for 5% to 8% of AMLs.

It is critical that the BM examiner remain vigilant for this disease for three reasons:

- There is a strong association with DIC, for which the patient should be immediately evaluated with PT, PTT and D-dimer studies. An undue delay in diagnosis in a patient with this complication can be deadly.
- Unique treatment consisting of initial administration of all-trans-retinoic acid (ATRA), which induces the cells to differentiate, followed by standard chemotherapeutic agents.
- If standard chemotherapy is given before the ATRA, then massive cell lysis with release of the cytoplasmic granules can result in a catastrophic exacerbation of the coagulopathy.

The prognosis is favorable if the patient survives the DIC and achieves remission. When the characteristic morphology is present, therapy is usually initiated immediately, before the diagnosis has been proven genetically.

The characteristic finding by FCM is the absence of HLA-DR expression. APL cells demonstrate CD33 positivity which is bright and homogeneous. They generally do not co-express CD15 and CD34. Some are CD117 positive. Hypogranular cases frequently express CD34 and CD2. Positivity for CD56 is associated with a poor outcome. Immunocytochemical staining using an antibody to the PML/RARA gene shows microgranular nuclear positivity with nucleolar exclusion.

FISH and PCR are more sensitive and rapid than cytogenetics for identifying t(15;17) and should be done immediately when this diagnosis is suspected. Sometimes complex variant translocations occur, some of which are cryptic. Most reference labs use FISH and PCR probes that detect both the traditional and variant abnormalities. Slight variations in morphology are seen with t(11;17) (q23;q12) which lack nuclear irregularities and Auer rods but show increased granularity. The t(11;17)(q23;q12) variant is resistant to ATRA. The FLT3 mutation is seen in about 40% and is associated with a high WBC count and microgranular blasts.

AML with t(9;11)(p22;q23); MLLT3-MLL

This leukemia most commonly demonstrates morphology consistent with AMML (M4), acute monocytic/monoblastic leukemia (M5) or, rarely, AML without (M1) or with (M2) maturation. It can be seen in all ages but predominates in children (11% of childhood leukemias). Numerous translocations of the mixed-lineage leukemia (MLL) gene have been identified which are related to both myeloid and lymphoblastic leukemias. The WHO recommends that the specific translocation be stated in the diagnosis. Approximately one-third of these abnormalities are cryptic. When associated with previous topoisomerase II inhibitor therapy, they should be diagnosed as therapy-related AML or AML with myelodysplasia-related changes. Some patients present with DIC and/or extramedullary monocytic sarcoma including involvement of the gingiva or skin. The prognosis is intermediate.

AML with t(6;9)(p23;q34); DEK-NUP214

This rare AML (about 1% of cases) is frequently associated with basophilia (>2% by definition) and multilineage dysplasia and often presents with cytopenias that are more prominent than generally seen with other AMLs. The morphology most commonly correlates with that of M4 (AMML) but may be that of any of the FAB types except M3 and M7. Some cases with this abnormality present with slightly less than 20% blasts and should be monitored closely for progression to AML. The overall prognosis is poor, and it has been proposed that allogeneic stem cell transplantation may offer better prospects for survival.

AML with inv(3)(q21;q26.2) or t(3;3)(q21;q26.2); RPN1-EVI1

This AML is sometimes associated with PB thrombocytosis and increased numbers of hypolobated megakaryocytes in the BM. Multilineage dysplasia is often present. Some patients present with hepatosplenomegaly. The prognosis is poor. When this genetic abnormality is acquired during the course of AML, it is a poor prognostic sign.

AML with t(1;22)(p13;q13), RBM15-MKL1

This leukemia occurs almost exclusively in infants (age <12 months) and has the morphology of acute megakaryoblastic leukemia (M7), from which it has been separated to become a unique entity in the 2008 revision of the WHO classification. Patients with acute megakaryoblastic leukemia with t(1;22) demonstrate extensive involvement of the abdominal organs by disease, resulting in hepatomegaly and/or splenomegaly and sometimes lymphadenopathy. Testicular and central nervous system involvement are not seen. These patients have a slightly lower PB WBC count and higher hemoglobin when compared to children with acute megakaryoblastic leukemia without t(1;22). The BM is usually hypocellular with reticulin fibrosis and a stromal pattern of involvement that may be easily mistaken for metastatic tumor. The disease course is aggressive with a poor prognosis.

AML with Normal Cytogenetics and Cytoplasmic/Mutated NPM

Mutation of the nucleophosmin 1 gene (NPM) is present in about one-third of all cases of AML and in about one-half of cases of AML-NK, which makes it the most common genetic abnormality in adult AML. The median age at presentation is 52 years, with a female predominance.

The NPM mutation cannot be detected by either standard cytogenetics or FISH and can only be identified using molecular techniques. Mutation of NPM can be inferred on IHC by localization of the protein in the cytoplasm of leukemic blasts (NPMc+), as opposed to the normal nuclear localization when un-mutated. Immunohistochemistry for cytoplasmic NPM may also aid in the evaluation for minimal residual disease. Whether or not IHC for NPM should be routinely done in all cases of AML-NK is still being debated.

Most cases are de novo AML, commonly with monocytic morphology, but any of the FAB subtypes may be seen except M3, M4eo and M7. Acute myeloid leukemias with mutated NPM are consistently negative for CD34, and they are not associated with other recurrent cytogenetic abnormalities.

The presence of NPMc⁺ is an independent favorable prognostic factor for achieving complete remission after induction chemotherapy, notably independent of FLT3 mutation status (see discussion of FLT3 below).

AML with Mutated CEBPA

The CCAAT/enhancer binding protein alpha (C/EBPα or CEBPA) mutation is found in 5% to 10% of cases of AML and in 15% of AML-NK and usually has the morphology of FAB type M1 or M2. Compared to AML-NK patients without mutated CEBPA, those with mutated CEBPA have higher hemoglobin, lower platelets and a higher PB WBC count. They are less likely to present with lymphadenopathy or other extramedullary disease. The rate of complete remission after induction chemotherapy is no different in patients with or without CEBPA mutation, but the duration of remission is longer in those with mutated CEBPA. The presence of the FLT3 mutation does not negatively affect the duration of remission of AML with mutated CEBPA (see discussion of FLT3 below). Overall survival (OS) is longer for patients with the CEBPA mutation. Familial cases of the rare **AML with maturation with eosinophilia (M2eo)** morphologic type that have the CEBPA mutation have been reported.

FLT3 Mutation

The FLT3 mutation is not one of the newly defined genetic abnormalities in the 2008 revision of the WHO classification but is a hot topic and is the target of novel therapies currently under investigation. Multiple types of somatic mutations may result in the constitutive activation of the fms-related tyrosine kinase 3 gene (FLT3) related to myeloid proliferation, which is independent of growth factors in murine models. The FLT3 mutation is not detected by standard cytogenetics and can be seen in cases of AML-NK. This mutation is found in about 30% of AMLs. These cases demonstrate cup-like nuclear invaginations that are also sometimes seen with the NPM mutation. FLT3 is generally associated with a poor prognosis.

AML with Myelodysplasia-Related Changes

These leukemias must demonstrate dysplasia in ≥50% of the cells of at least two different cell lines or possess MDS-related cytogenetic abnormalities or occur in the setting of a previously established diagnosis of MDS. Furthermore, the dysplasia must be identified prior to treatment since chemotherapy can cause dysplastic changes in the erythroid and megakaryocytic cells. The recurring genetic abnormalities that define specific entities in the WHO classification must be absent. Patients often present with severe pancytopenia. The prognosis is inferior to that of equivalent leukemias lacking myelodysplasia-related changes.

Therapy-Related Myeloid Neoplasms

This group includes therapy-related AML (t-AML), MDS (t-MDS) and myelodysplastic/myeloproliferative neoplasms (t-MDS/MPN). These leukemias were previously separated into those related to topoisomerase II inhibitor therapy and those associated with alkylating agents or radiation therapy because of differences in the latency periods, morphology, genetics and prognosis. The new WHO classification does not make this distinction, since it has been recognized

Table 7.12. Key features related to specific treatments of therapy-related myeloid neoplasms

	Alkylating Agent/Radiation Therapy	Topoisomerase II Inhibitor Therapy
Latent period	2-11 years	<1-3 years
	May evolve from MDS	No preexisting MDS
Morphology	Most commonly AML with maturation, AMML, and acute erythroid leukemias	Most commonly acute monocytic/monoblastic leukemia or AMML
	Multilineage dysplasia	
	Severe pancytopenia common	
Genetics	Del-5/5q-,	11q23 translocations
	-7/7q-	
	Frequent complex (>3 recurring) abnormalities	
Prognosis	Resistant to therapy and poor	Similar to that of equivalent AML not related to therapy

that most patients have been given polytherapy and there is much overlap in the above characteristics. These therapies have in common the fact that they induce permanent DNA damage, which is theorized to be the first step in the development of most malignancies. The diagnosis may be placed in one of the other WHO-defined categories if the appropriate criteria are met, and then given the qualifier, "therapy-related." It is still clinically useful to recognize some of the key features related to specific treatments (Table 7.12). The prognosis is generally poor but is related to the genetic abnormalities that are present and the underlying disease for which the patient was receiving therapy.

Myeloid Proliferations Related to Down Syndrome

Down syndrome (DS) is associated with an increased risk for acute leukemia, predominantly AML, most commonly acute megakaryoblastic leukemia. In some patients there is spontaneous remission of the leukemic findings over several weeks, which is then referred to as **transient abnormal myelopoiesis.** Acute megakaryoblastic leukemia develops subsequently in 20% to 30% in 1 to 3 years, referred to as **myeloid leukemia associated with Down syndrome.** These diseases have unusual blast morphology with round or irregular nuclei and basophilic cytoplasm that sometimes contains coarse basophil-like granules and/or demonstrates pseudopods. Erythroid dysplasia is common, and the PB may be involved by basophilia. The acute leukemia and transient disease are not morphologically distinguishable from one another. Important features are summarized in Table 7.13.

Table 7.13. Important features of myeloid proliferations related to Down syndrome

FAB	None but often similar to M7
Lineage	Myeloid and megakaryocytic
Definition	AML, usually megakaryocytic, occurring in neonatal period in patients with Down syndrome
	Cases that undergo spontaneous remission designated as transient abnormal myelopoiesis
	Usually demonstrate >30% blasts
Frequency	10% of DS newborns for transient abnormal myelopoiesis
	1-2% of DS children develop AML before age 5
Cytochemistry	Negative for MPO, SBB, TdT
	Some show granular PAS positivity
FCM	Positive for CD36 and CD41 and/or CD61
	Less often positive for CD42
	May be positive for CD13, CD33, or CD7
	Often negative for CD34, CD45, HLA-DR
IHC	Positive for FVIII, CD61
Genetics	Trisomy 21
	GATA1 gene mutations
	Trisomy 8 and, rarely, monosomy 7 of leukemic cells
	Negative for t(1;22)
Special features	Promegakaryocytes and micromegakaryocytes sometimes seen in PB
Prognosis	Most undergo spontaneous remission in 1-3 months, but 20%-30% develop AML
	Recurrences respond to chemotherapy but have poor prognosis

Blastic Plasmacytoid Dendritic Cell Neoplasm

This rare neoplasm of blastic plasmacytoid dendritic cells (BPDC) typically presents as multiple skin lesions and sometimes with lymphadenopathy (20%), but with minimal involvement of the PB and BM that subsequently progresses. Approximately 10% to 20% of cases are associated with or develop AMML, which must be differentiated from AMML with massive nodal or extranodal involvement by mature, CD56 positive BPDCs. The tumor cells are of intermediate size with fine chromatin, one to multiple nucleoli, nuclear irregularities and scant gray-blue cytoplasm on Wright-Giemsa staining. The PB and BM aspirate

Table 7.14. Immunophenotype of BPDC neoplasm

- Positive for CD4,CD43, CD45RA, CD56
- BPDC associated antigens CD123, BDCA-2/CD303, TCL1, CLA, MxA
- Positive for CD68 in 50% of cases
- Commonly express CD7, CD33
- Rarely express CD2, CD36, CD38
- Negative for lysozyme, MPO, CD34, CD117
- Positive for TdT in 33% of cases seen in 10%-80% of cells

smears may contain tumor cells that show cytoplasmic vacuolizations aligned along the cell membrane and pseudopodia. The BM biopsy may demonstrate interstitial involvement that can only be identified by IHC or, alternatively, be massively effaced by tumor. The immunophenotype is given in Table 7.14 and specific recurrent cytogenetic abnormalities in Table 7.15.

The prognosis is poor (median survival 12 to 14 months), usually showing an initial response to chemotherapy followed by an inevitable relapse.

Examination of the Bone Marrow Biopsy Specimen, Special Considerations

Most of the previous discussion of AML refers to the BM aspirate findings because they are the mainstay of AML diagnosis and classification. A few words should be said about two special circumstances: when an adequate BM aspirate sample is not available and when the BM biopsy is hypocellular.

When an aspirate is not available: Occasionally, obtaining an adequate BM aspirate specimen in AML is not possible because the marrow is tightly packed with malignant cells, there is fibrosis related to the AML or there is scarring from previous bone marrow sampling. In these cases, PB evaluation as well as morphology and IHC of the BM biopsy become critical.

The PB can save the day if sufficient numbers of blasts are present, in which case it can be submitted for cytochemistry, FCM, cytogenetics, FISH, PCR or any other studies which would normally be done on the BM aspirate. If there are ≥20% blasts present in the PB or a cytogenetic abnormality defined in low blast count AML is identified, then a diagnosis of AML can be made. Otherwise, BM biopsy morphology and IHC must be relied upon heavily.

Table 7.15. Recurring cytogenetic abnormalities in BPDC neoplasm

Cytogenetic Abnormality	Occurrence
5q21 or 5q34	72%
12p13	64%
13q13-21	64%
6q23-qter	50%
15q	43%
Loss of chromosome 9	28%

Table 7.16. IHC stains useful in BM biopsy evaluation

Cell Type	IHC Stain
Hematologic precursors	CD34
Myeloid	MPO
Promyelocytes	MPO
Monocytic	CD68, CD163, Lysozyme
Erythroid	Gly A, Hgb A, CD71
Megakaryocytic	FVIII, CD61, CD41
Mast cells	CD117, tryptase

When the BM biopsy contains a large population of immature cells, the lineage can usually be determined with IHC. Immunohistochemical markers that are helpful for the identification of cell lineage have been previously mentioned, but a summary is given in Table 7.16. It becomes more problematic when the blast population is morphologically subtle or when there is monocytic differentiation. A CD34 stain may be positive in ≥20% of the cells, in which case a diagnosis of AML can be made. It is less helpful if the CD34 stain is positive in <20%. Remember that all CD34 positive cells are blasts, but not all blasts are CD34 positive. Lymphoblasts may be positive for CD34, but myeloblasts are usually MPO positive and negative for TdT. In the case of monocytic differentiation a CD68, CD163 or lysozyme stain will be positive.

Touch preparations of the BM biopsy can be useful when there is not an adequate BM aspirate specimen. Because the blasts are not subjected to fixation or decalcification, and the slides are Wright-Giemsa stained, the morphology is similar to that seen in an aspirate specimen. Touch preparations are superior to the BM biopsy for the identification of dysplasia as well as the differentiation of blasts from other immature myeloid cells (myelocytes, promyelocytes). Touch preparations should be done at the time of BM sampling whenever the operator feels the aspirate may be suboptimal.

Flow cytometry: The BM biopsy specimen is usually considered inappropriate for FCM; however, non-decalcified, morcellated fragments may be flowed. The pathologist should ensure adequate tissue remains for morphologic examination.

Classification of AML based solely on the biopsy: In many cases, AML cannot be classified based on examination of the BM biopsy without an aspirate, and in some cases a diagnosis of AML cannot be made. In these situations correlation with the clinical findings and use of clinical judgment should be employed as best as possible. Do not fall into the trap of over-interpreting limited data. Sometimes a specific diagnosis cannot be made, and the pathologist must give a descriptive diagnosis and ask for re-sampling of the BM.

Hypoplastic AML: As with MDS, the BM in AML can rarely be hypocellular. These cases, referred to as hypocellular AML or hypoplastic AML, are not recognized as a separate disease entity by the WHO classification system. The major differential diagnosis of hypoplastic AML is aplastic anemia (AA). In AA the BM is usually less cellular than with hypoplastic AML and demonstrates only sparse interstitial lymphocytes and plasma cells with rare maturing hematopoietic cells. In hypoplastic AML the BM is usually more cellular than with AA or simply lacks hypercellularity (i.e., normocellular for the patient's age) and invariably contains a significant blast population which can be highlighted with appropriate IHC stains. A CD34 stain, if positive, can be particularly helpful. It is always a good idea to do a CD20 stain on hypocellular BM biopsy specimens to rule out hairy cell leukemia, which is a treatable disease with an excellent prognosis.

Evaluation of Post-Treatment Bone Marrow in AML (Figs. 7.14, 7.15)

There are two primary reasons for examination of the BM after treatment:
- To assess for residual AML
- To evaluate BM recovery following myelosuppression

Evaluation of remission: "Complete remission" is defined as <5% blasts in the setting of maturation of all myeloid cell lines in a BM with ≥20% cellularity. Evidence of myeloid maturation includes a platelet count >100 x 10^9/L and an absolute neutrophil count (ANC) >1.5 x 10^9/L. Anemia is acceptable if the other conditions are met. Furthermore, there can be no Auer rods seen and no blasts in the PB. These findings should be present for more than four weeks. The term "partial remission" is sometimes used in clinical trials and allows >5% blasts or <5% blasts with Auer rods. In the community hospital setting, post-treatment BM samples are usually taken to evaluate for complete remission. If the patient is a participant in a research protocol, then definitions for remission will be provided or slides are submitted to the sponsoring institution for examination. For practical purposes the BM is usually reported as being "in remission" when the blast count is <5%. FCM can be used to identify small numbers of residual leukemic cells (minimal residual disease or MRD) and provides a reassuring cross-check of the blast count in these typically hypocellular specimens. In some cases the blast count in the aspirate is <5%, but clusters of blasts are seen on the biopsy specimen. If these clusters are CD34 positive, then they probably represent residual disease. However, discerning myeloblasts from slightly more mature regenerating cells (promyelocytes and myelocytes) can be difficult. Most such cases should have the BM re-sampled.

Evaluation for BM recovery following myelosuppression: In some cases the clinician performs post-therapy BM sampling because the PB cell counts are not recovering. Recovery occurs in the following sequence: erythroid cells followed and overlapping with myeloid cells and, lastly, megakaryocytes. The average time for complete recovery is 60 days and should take no longer than 100 days. Hematogones (see description in ALL chapter) may appear in adults and children around day 40 and peak on day 60. Recovery of cell counts may be delayed for a variety of reasons, but the clinician is primarily interested in ruling out residual leukemia and myelodysplasia.

Megaloblastic changes: Therapy with anti-metabolite drugs is part of the common regimens used for AML and may induce megaloblastic changes. These changes can be indistinguishable from myelodysplasia; therefore, myelodysplasia should only be diagnosed if the therapy is remote.

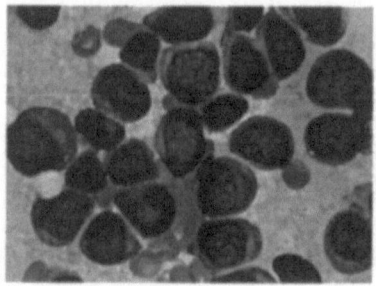

Figure 7.1. AML, without maturation (BM aspirate, 500x).

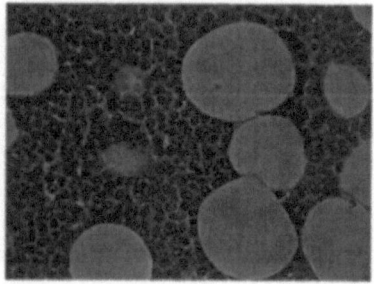

Figure 7.2. AML, without maturation (BM biopsy, 400x).

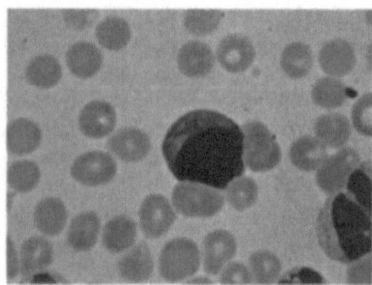

Figure 7.3. Myeloblast with Auer rods (PD, 500x).

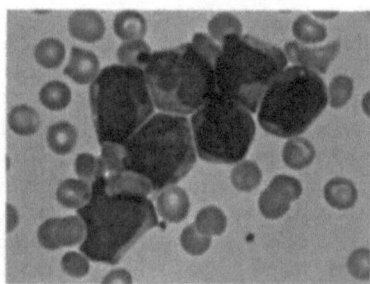

Figure 7.4. Myeloblasts with marked nuclear irregularities (PB, 500x).

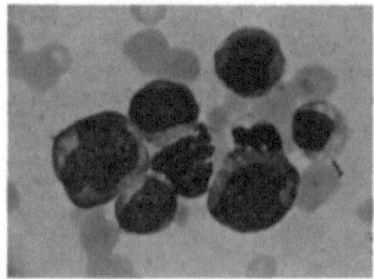

Figure 7.5. Acute monoblastic leukemia (BM aspirate, 500x).

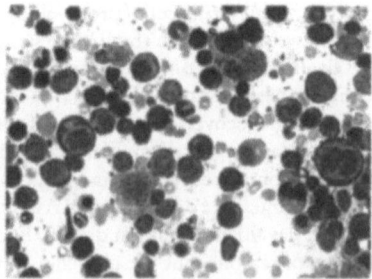

Figure 7.6. Acute erythroid leukemia (BM aspirate, 200x) (photo courtesy of Dr. John Lazarchick, Medical University of South Carolina).

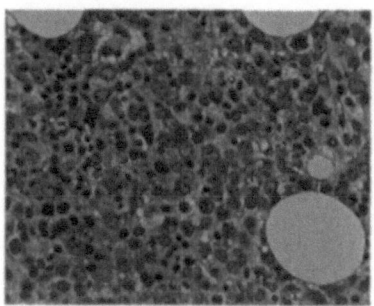

Figure 7.7. Acute erythroid leukemia (BM biopsy, 200x) (photo courtesy of Dr. John Lazarchick, Medical University of South Carolina).

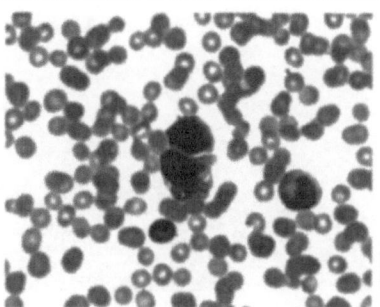

Figure 7.8. Acute megakaryoblastic leukemia (BM aspirate, 400x) (photo courtesy of Dr. John Lazarchick, Medical University of South Carolina).

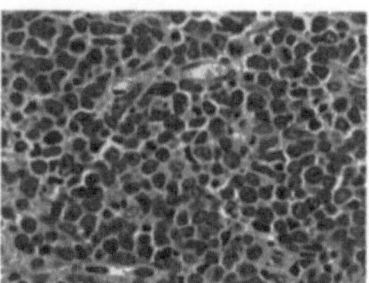

Figure 7.9. Acute megakaryoblastic leukemia (BM biopsy, 200x) (photo courtesy of Dr. John Lazarchick, Medical University of South Carolina).

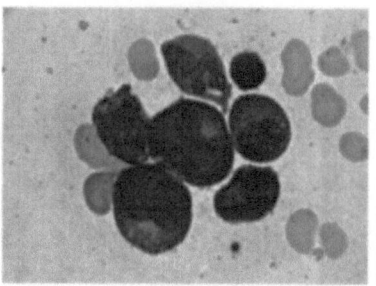

Figure 7.10. Acute myelomonocytic leukemia with eosinophilia (BM aspirate, 500x).

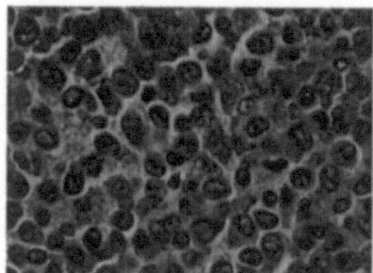

Figure 7.11. Acute myelomonocytic leukemia with eosinophilia (BM biopsy, 400x).

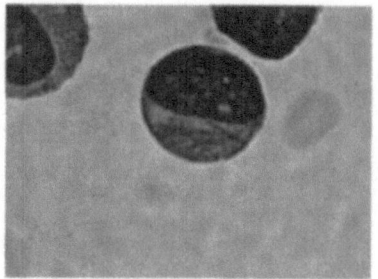

Figure 7.12. Acute promyelocytic leukemia, faggot cell (BM aspirate, 1000x).

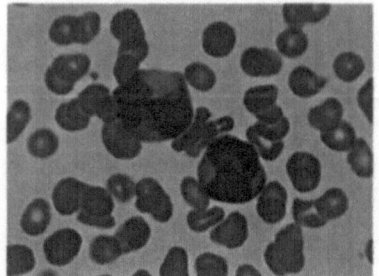

Figure 7.13. Acute promyelocytic leukemia, hypogranular variant (PB, 500x).

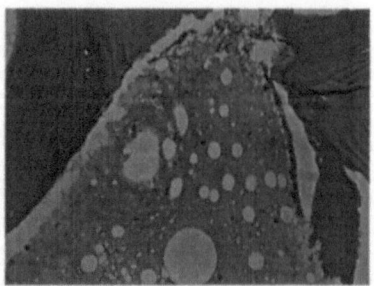

Figure 7.14. Serous atrophy following induction chemotherapy for AML (BM biopsy, 100x).

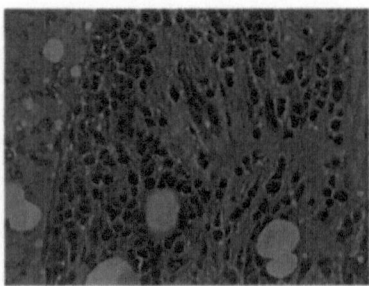

Figure 7.15. Treatment failure; residual AML in background of serous atrophy (BM biopsy, 100x).

CHAPTER 8

Acute Lymphoblastic Leukemia

Introduction

Acute lymphoblastic leukemias (ALLs) are clonal malignancies of immature cells of lymphocytic lineage that involve the bone marrow (BM), peripheral blood (PB) and, in some cases, extramedullary sites. Approximately 80% of ALLs occur in children, which is the inverse of acute myeloid leukemia (AML), and an estimated 3,200 new cases are diagnosed in the United States each year. Acute lymphoblastic leukemia is the most common cancer in children.

Clinical presentation and sites of involvement: The clinical features of ALL are usually related to cytopenia(s) such as fatigue and pallor due to anemia, bruising and petechiae secondary to thrombocytopenia, and infection and fever resulting from neutropenia. Other clinical findings are caused by the massive expansion of tumor cells in the PB leading to thrombotic complications, or in the BM causing bone or joint pain, which is a particularly common presenting complaint from children. Small numbers of patients have extramedullary involvement of the central nervous system (CNS), skin, lymph nodes (LNs), gonads, kidney or other sites. Disseminated intravascular coagulation (DIC) may result from cytokine release into the circulation.

Terminology of leukemia versus lymphoma: ALL is divided into two groups:
- Precursor B lymphoblastic leukemia/lymphoma (B-ALL/LBL)
- Precursor T lymphoblastic leukemia/lymphoma (T-ALL/LBL)

The lymphocytic lineage is based on the immunophenotype as determined with flow cytometry (FCM). Approximately 80% of cases are B-ALL. The terms "leukemia" and "lymphoma" are based on the dominant presentation of the tumor. If the patient presents with a mass and ≤25% lymphoblasts in the BM, then the designation "lymphoma" is used. In contrast to AML, terminology from the old French American British (FAB) classification system is seldom used and will not be discussed. Hereafter, the terms B-ALL or T-ALL will be used to refer to both ALL and LBL unless otherwise stated.

Work-up of suspected cases of ALL: The same general principles described for the evaluation of patients with suspected AML apply to the work up of ALL.
1. Abnormal cells in the PB and/or BM are identified as blasts.
2. Lineage and maturity of the blasts are determined by FCM.
3. Genetic studies are performed to further categorize the disease.

Lymphoblast morphology: When examined in the PB or BM aspirate, lymphoblasts are small to medium in size, about twice the size of normal lymphocytes, and have moderately condensed reticular chromatin, inconspicuous nucleoli and scant moderately basophilic cytoplasm that may contain vacuoles or granules.

Bone Marrow: A Practical Manual, by Daniel A. Cherry and Tomislav M. Jelic.
©2011 Landes Bioscience.

Lymphoblasts never contain Auer rods but, otherwise, cannot be reliably distinguished from myeloblasts based on morphology alone. Likewise, B and T-cell differentiation cannot be determined by morphology. In BM biopsy sections, lymphoblasts are usually present in a diffuse pattern with moderately condensed nuclear chromatin, inconspicuous nucleoli and barely-discernable cytoplasm. The low power appearance is commonly described as "uniform."

Differentiation from hematogones: Hematogones are normal B lymphocyte precursor cells that are commonly present in increased numbers in pediatric patients and are similar in appearance to lymphoblasts. These cells may comprise 5% to more than 50% of the cells in the BM differential count in children and adults with

- Non-leukemia neoplasms
- BM regeneration following ablative therapy
- AIDS
- Cytopenias due to immune and congenital causes

Hematogones are small with oval or slightly indented nuclei, smooth homogeneous chromatin and absent or barely discernable cytoplasm. Hematogones can be reliably distinguished from B-lymphoblasts by experienced flow cytometrists based on their expression of a spectrum of antigens that typify normal B-cell evolution, as opposed to the incomplete and aberrant expression of antigen maturation seen with neoplastic lymphoblasts. Immunophenotypic differentiation from T lymphoblasts poses little difficulty, since hematogones are of B lineage.

Characterization of lymphoblasts using FCM can differentiate B lymphoblasts, T lymphoblasts, and myeloblasts as well as distinguish lymphoblasts from more mature neoplastic lymphoid cells such as those of blastoid mantle cell lymphoma. The immunophenotypic features of ALL are discussed in more detail below.

Cytochemistry is currently used far less frequently than in the past because FCM is superior in almost every way. The main utility of cytochemistry lies in differentiating ALL from AML in the setting of a critically ill patient in which delaying treatment while awaiting FCM results from a distant reference laboratory is unacceptable. Cytochemistry is discussed more fully in the separate sections on B-ALL and T-ALL.

B-ALL/LBL

B-ALL is predominantly a disease of children, with approximately 75% of cases occurring before the age of six years. B-LBL comprises approximately 10% of cases of lymphoblastic lymphoma.

B-ALL may occur in extramedullary sites:

- CNS
- Spleen
- Lymph nodes (LNs)
- Liver
- Gonads

B-LBL most commonly involves

- Skin
- Bone
- Soft tissues
- LNs

Table 8.1. Atypical morphologic features seen with B-ALL

Morphologic Feature	Comments
Cytoplasmic vacuolization	Common finding; composed of glycogen, lipid or organelles
Cytoplasmic granules	Negative for MPO but may stain faintly with PAS or SBB; more common in children with Down syndrome and cases with t(9;22)
Pancytopenia and BM hypoplasia	Rare; usually followed by overt leukemia weeks or a few months later
Mature-appearing blasts	Difficult to differentiate from other lymphocytic neoplasms with mature morphology (chronic lymphocytic leukemia, mantle cell lymphoma, marginal zone lymphoma)
BM necrosis	Always suggestive of malignancy; also occurs in Burkitt lymphoma and metastatic small round blue cell tumors*; may require repeated biopsies to obtain adequate morphology for diagnosis
Hypereosinophilia	Rare; associated with t(5;14)(q31;q32); IL-3 levels elevated; typically present with hypereosinophilic syndrome (pulmonary infiltrates, cardiac symptoms, CSF eosinophilia) prior to leukemia; also seen in rare instances with T-ALL

MPO: myeloperoxidase; PAS: periodic acid Schiff; SBB: Sudan black B
*Neuroblastoma, Ewing sarcoma, medulloblastoma

In contrast to T-LBL, B-LBL rarely occurs in the mediastinum. A variety of atypical morphologic features may be seen with B-ALL (Table 8.1).

Cytochemistry of B-ALL: The primary utility of cytochemistry is for ruling out AML in the early evaluation of patients when FCM must be sent to a remote laboratory and in some high volume laboratories for rapid differentiation from AML. B-lymphoblasts are always negative for MPO and may show light-gray staining of cytoplasmic granules with Sudan black B (SBB), as opposed to the intense staining seen in myeloblasts. B-lymphoblasts may show periodic acid Schiff (PAS) and nonspecific enolase (NSE) reactivity.

Immunophenotype: B-ALL demonstrates variable patterns of antigen expression as demonstrated on immunophenotyping with FCM or immunohistochemistry (IHC). For the purpose of understanding these immunophenotypic patterns, it is useful to divide the antigens into those that are lineage specific, nonspecific, aberrant (Table 8.2) and those that infer stage of maturation. Antigens associated with maturation stage and clinical associations are given in Table 8.3. It should be kept in mind that the term "specific" in this context is relative, as exceptions have been found in all instances.

HLA-DR and CD19 are expressed in almost all (98%) cases of B-ALL and rarely in T-ALL, but may be seen in AML. Terminal deoxynucleotidyl transferase

Table 8.2. Commonly expressed antigens in B-ALL

B lymphocyte antigens	CD19, CD22, CD79a, CD24, CD10, CD9
Lineage nonspecific antigens	CD34, TdT, HLA-DR, CD38, CD45
Aberrant (myeloid) antigens	CD13, CD33, CD15

(TdT) is seen in virtually all cases of B-ALL and T-ALL but is present in only 10% of cases of AML.

Other antigen patterns that are of interest include expression of CD10, which is seen in most children but in a minority of adults and in only about one-third of cases of T-ALL. TdT is seen in B-ALL and T-ALL but not in neoplasms of more mature lymphoid cells. Almost all B-ALL cases lack surface immunoglobulin, which is present in the majority of more mature B-cell neoplasms.

Genetic abnormalities can be identified in nearly all cases of B-ALL and include numerical as well as structural changes. This information should be sought and reported in all cases, since many of these abnormalities are of significance to prognosis and treatment. The WHO recognizes the importance of several genetic alterations and sorts them into favorable and unfavorable prognostic groups. Structural genetic abnormalities are listed in Table 8.4, with those recognized by the WHO presented in bold print. Numerical abnormalities are depicted in Table 8.5 in a similar fashion. In general, the prognostic significance of various genetic profiles is less well characterized in adults than in children.

Special studies: Clonal immunoglobulin heavy chain (IgH) gene rearrangements can be detected by PCR, FISH or Southern blot in most cases of B-ALL but are nonspecific since they are also present in many other B-lymphocytic neoplasms. This finding is of little utility in the evaluation of B-ALL.

Table 8.3. Antigen patterns associated with maturation stage in B-ALL

Maturation Stage	Antigens	Associations
Early precursor B-ALL	TdT, CD34 and CD19, and/or CD22, and/or CD79a	Often seen with abnormalities of chromosome 11q23 (bad prognosis), which also often express myeloid antigens (CD15, CD13, CD33)
Common precursor B-ALL	Defined by expression of CD10; also usually express TdT, CD19, HLA-DR, CD34	Most frequent type of B-ALL in both children and adults; cases fall within both favorable and unfavorable cytogenetic groupings
Pre-B-ALL	Defined by expression of cytoplasmic μ (cyt-μ); also usually express CD19, CD22, CD10, TdT	20% have t(1;19)(q23;q13)

Table 8.4. Structural genetic abnormalities in B-ALL, with those recognized by the WHO in bold print

Abnormality	Gene	Prognostic Group	Immuno-phenotype	Special Features
t(9;22)(q34; q11.2)	BCR/ ABL	Un-favorable	Variable	High WBC at presentation and older age; cytoplasmic granules more common; CNS involvement and organomegaly more common
t(4;11) (q21;q23)	AF4/ MLL	Un-favorable	Early precursor B-ALL; CD10 negative, positive for TdT, CD34, HLA-DR; commonly express myeloid antigens CD15, CD13, CD33	MLL associated with markedly elevated WBC; increased frequency of CNS involvement; more common in children
t(1;19)(q23; p13.3)	PBX/ E2A	Un-favorable	CD19, CD10, CD9; negative for CD34; CD20 negative or decreased	Some studies show increased frequency in African Americans, high WBC and increased CNS involvement in African Americans
t(12;21) (p13;q22)	TEL/ AML1	Favorable	Common precursor B-ALL; CD10, HLA-DR; negative for CD10, CD20; many express myeloid antigens CD13, CD33	Cryptic (not detected by standard cytogenetics but by FISH, PCR or Southern blot)
t(17;19) (q21-22; p13)	E2A/ HLF	Un-favorable	Variable	More common in adolescents; increased presentation with hypercalcemia and DIC; very rare (<1% of B-ALL)
t(5;14) (q31;q32)	IL-3/ IgH	Un-favorable	Variable	Hypereosinophilia
Del(6q)	Not cloned	Does not affect prognosis	Variable	Most are associated with other cytogenetic abnormalities
9p abnormalities	Not applicable	Un-favorable	Variable	Usually associated with complex karyotype (>3 abnormalities)

Table 8.5. Numerical genetic abnormalities in B-ALL, with those recognized by the WHO in bold print

Abnormality	Prognostic Group	Comments
Hyperdiploidy >50 Chromosomes	Favorable	Correlates with DNA index of 1.16; duplication of two or more of chromosomes 4, 10, or 18 are especially favorable; favorable outcome linked to response to anti-metabolite therapy
Hyperdiploidy 47-50 chromosomes	Intermediate	Approximately 50% have associated structural abnormalities
Hypodiploidy	Unfavorable	Unfavorable in adults but intermediate in children; usually have 45 chromosomes; chromosome 20 commonly lost in children; near-haploid has particularly bad outcome
Diploid (not abnormal, but included for completeness)	Intermediate	Occurs in 30% of B-ALL in adults and 10% in children

Prognosis: The prognosis of B-ALL is good, with complete remission rates of 95% in children and 60-85% in adults. In addition to genetic findings, other factors are prognostically important (Table 8.6). It should be noted that some of these are not statistically independent but are related to genetic factors or vice versa.

Evaluation for relapse: Relapse of B-ALL is usually identified in the BM of patients with decreasing PB cell counts. BM relapse may be difficult to identify because of small numbers of cells, focality of disease (which is more common with relapse than at initial presentation), and difficulties with differentiation from

Table 8.6. Summary of prognostic factors in B-ALL

Prognostic Factor	Favorable	Unfavorable
Age	4-10 years	<1 or >10 years
WBC	Normal or decreased	>50 x 10⁹/L
Initial response to therapy	BM disease-free by day 14	Residual BM disease at day 14
Relapse after remission obtained	Absent	Present
CD10 expression	Positive	Negative
Genetic abnormalities	t(12;21)(p13;q22), Hyperdiploidy >50 chromosomes	t(9;22)(q34;q11.2), t(4;11)(q21;q23), t(1;19) (q23;p13.3), hypodiploidy

regenerative changes. IHC for TdT and CD34 may be helpful, but FCM and molecular studies are far more sensitive. Material for these studies should be sent with any BM done to evaluate for relapse.

Occasionally relapse of B-ALL occurs in extramedullary sites. The most common is the CNS. The second most common, in males, is the testicle. Other sites include the skin, eye and LNs. In some cases of extramedullary relapse, occult BM involvement may be found if carefully searched for. FCM is a highly effective method for identifying small numbers of cells in the CNS and differentiating them from the large atypical non-neoplastic lymphocytes commonly encountered in CSF. Appropriate samples for FCM should be collected with all lumbar punctures performed on these patients. Keep in mind that a relapse of leukemia may be treatment-related AML, particularly when topoisomerase II inhibitor drugs have been used.

T-ALL/LBL

There are many similarities between T-ALL and B-ALL. The following section is focused on the specific differences.

Unlike B-ALL, T-ALL is more common in older children. T-ALL is the most common malignancy in adolescent males. T-ALL accounts for 85–90% of lymphoblastic lymphomas.

Compared to B-ALL, the presenting WBC with T-ALL is generally much higher, with a median of about 100×10^9/L, and the cytopenias tend to be less pronounced because of the relative sparing of normal hematopoietic marrow.

Approximately 50% of patients present with a mediastinal mass. Other extramedullary sites of involvement are as follows:

- LNs (other than mediastinal)
- Skin
- Liver
- Spleen
- Waldeyer's ring
- CNS
- Gonads

Morphology: T lymphoblasts are not morphologically discernable from B lymphoblasts. In many cases there is a variably-sized, distinct population of small blasts with marked nuclear hyperchromasia, prominent nuclear irregularities and scant to no cytoplasm. This finding is rare in B-ALL. These small cells may also be appreciated in BM sections. Otherwise, the morphology of T-ALL in the BM sections is similar to that of B-ALL, with the exception of a much higher mitotic rate in the former.

Morphologic variants similar to those observed with some cases of B-ALL may also be seen with T-ALL, including the following:

- Cytoplasmic granules
- Mature-appearing blasts
- Associated hypereosinophilia

There is a rare and unusual association of the t(8;13)(p11.2;q11-22) chromosomal abnormality with T-ALL, eosinophilia and myeloid hyperplasia. Some of these patients subsequently develop AML or a myelodysplastic syndrome (MDS).

Cytochemistry: T-ALL demonstrates focal positivity for acid phosphatase. Negativity for myeloperoxidase is more relevant because of the utility of this finding to rule out AML in the early evaluation of these patients.

Immunophenotype: In general, the immunophenotypic and genetic profiles in cases of T-ALL are less complex and do not have the same prognostic significance or treatment implications compared to B-ALL. T lymphoblasts variably express T-associated antigens, including CD1a, CD2, CD3, CD4, CD5, CD7 and CD8. Only CD3 is lineage specific and, along with CD7, is the most commonly expressed antigen. Most (>90%) express TdT, which is useful for differentiating T-ALL from peripheral T-cell lymphoma. CD4 and CD8 are commonly co-expressed. Approximately 20% of cases demonstrate CD10 positivity. The B-cell marker CD79a and the myeloid antigens CD13 and CD33 may be identified. CD117 is rarely present. The maturation stage may be inferred by the immunophenotype, with CD2, CD3 and CD7 seen in early stages, CD5 and CD1a in mid-stage and membranous CD3 in late stage. These antigen groupings lack the clinical correlations seen with similar maturation groups in B-ALL.

Genetics: Recurrent cytogenetic abnormalities have been identified in T-ALL, including those of MYC (8q24.1), TAL1 (1p32), RBTN1 (11p15), RBTN2 (11p13) and HOX (10q24) but do not carry prognostic importance.

Special studies (differentiation from thymocytes): Clonal T-cell receptor (TCR) gene rearrangements as identified by FISH, PCR or Southern blot are present in most cases. These studies may be invaluable when struggling with the differentiation of T-lymphoblasts from normal thymocytes in the evaluation of tissue from a mediastinal mass. Remember that normal thymocytes are positive for TdT.

Prognosis: Clinical risk factors for T-ALL are similar to those of B-ALL. There is an association of increased WBC and age over 15 years with an unfavorable outcome. The prognosis of T-ALL treated with intense regimens is comparable to that of B-ALL.

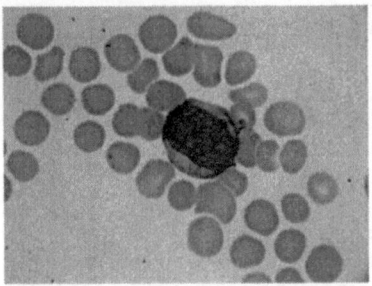

Figure 8.1. Lymphoblast with typical appearance (PB, 500x).

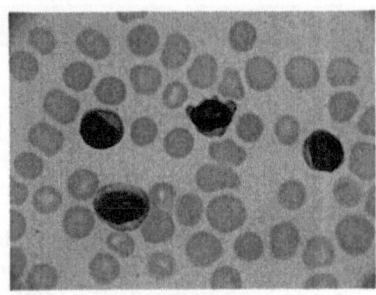

Figure 8.2. ALL with mature-appearing lymphoblasts (PB,500x).

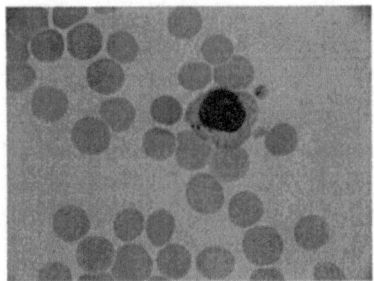

Figure 8.3. Lymphoblast with cytoplasmic granules (PB, 500x).

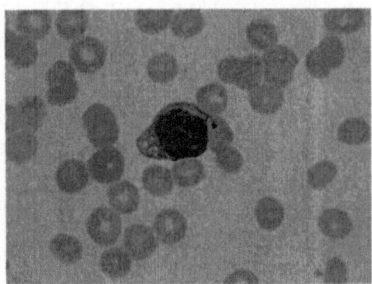

Figure 8.4. Lymphoblast with cytoplasmic vacuoles (PB, 500x).

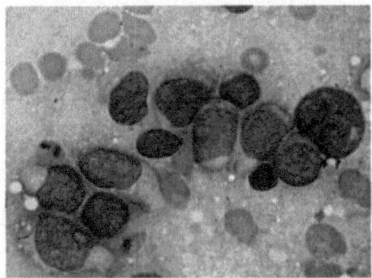

Figure 8.5. ALL (BM aspirate, 500x).

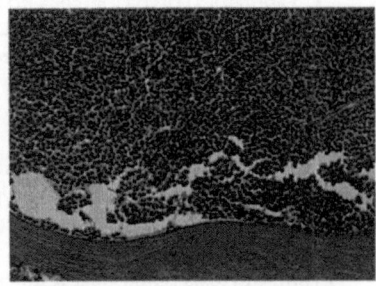

Figure 8.6. ALL (BM biopsy, 500x).

CHAPTER 9

Mature B, T and NK Cell Neoplasms

Introduction

This chapter serves as a gathering place for a diverse group of mature B, T and NK-cell neoplasms that have prominent involvement of the peripheral blood (PB) and bone marrow (BM). Lymphomas predominantly arise in the lymph nodes (LNs) but can arise in other organs (extranodal lymphomas). Almost all lymphomas may involve the PB and/or BM, especially when they are advanced. A general discussion of these lymphomas can be found in the chapter on evaluation of the BM for lymphoma. In the WHO classification scheme, the neoplasms discussed in this chapter are designated under the categories of "Mature B-Cell Neoplasms" and "Mature T-Cell and NK-Cell Neoplasms" (Table 9.1). Chronic lymphocytic leukemia (CLL) is the most common of these, while the others are rare. All primarily affect adults.

Approach to the Diagnosis of Mature Lymphocytic Leukemias

The pathologist's tasks in these cases are to decide if there is sufficient reason to suspect a neoplasm, initiate work-up of suspicious cases, interpret and integrate the information from the work-up into a diagnosis, and obtain additional information relevant to treatment.

The most difficult of these tasks is probably the first. Many of the mature lymphocytic neoplasms can be quite subtle in their early course, and the key to recognizing them is to constantly keep them in mind. Unlike many of the other diseases in this book, evaluation of the PB (as opposed to the BM) plays the primary role in the diagnosis. BM sampling is probably not necessary in most instances. These neoplasms usually come to the attention of the pathologist by way of the PB, which may be sent to the pathologist for a variety of reasons:

- Laboratory criteria for reflex PB review (markedly elevated WBC, increased atypical lymphocytes, cytopenias)
- At the request of a medical technologist to identify unrecognizable or severely atypical cells
- At the request of a clinician who suspects a hematologic disease on clinical grounds

Cytopenias: Neutropenia is a vital clue when seen in the company of large granular lymphocytes (LGLs) or atypical lymphocytes (ATLs), and hepatosplenomegaly or lymphadenopathy. Monocytopenia is characteristic of hairy cell leukemia (HCL).

Lymphocytosis is nonspecific unless extremely high. Neoplastic lymphocytes can be few in number in some of these diseases, such as HCL.

Bone Marrow: A Practical Manual, by Daniel A. Cherry and Tomislav M. Jelic.
©2011 Landes Bioscience.

Table 9.1. Mature B, T and NK cell neoplasms

- Chronic lymphocytic leukemia (CLL)
- B-cell prolymphocytic leukemia (B-PLL)
- Hairy cell leukemia (HCL)
- T-cell prolymphocytic leukemia (T-PLL)
- T-cell large granular lymphocytic leukemia (T-LGL)
- Indolent large granular NK-cell lymphoproliferative disorder
- Aggressive NK-cell leukemia
- Adult T-cell leukemia/lymphoma (ATLL)

Lymphocyte morphology shows much overlap among the various neoplasms and reactive conditions. Neoplastic lymphocytes may have normal morphology (CLL), very subtle morphologic changes (HCL) or morphologic changes that are within the spectrum of what can be seen in reactive processes (LGL leukemias).

The pathologist should review any available clinical information at the time of PB examination because there may be important evidence for neoplasia.

Hepatomegaly and/or splenomegaly are important findings favoring neoplasia since they are not typically seen with reactive disorders.

Lymphadenopathy, when limited or localized, may be seen with either reactive or neoplastic conditions. When diffuse or generalized, lymphadenopathy favors neoplasia. Persistence of lymphadenopathy for longer than three months favors neoplasia.

Rheumatoid arthritis has an association with T-LGL leukemia.

Constitutional symptoms (fever, weight loss, night sweats) are nonspecific and may occur with many reactive and neoplastic processes.

Flow cytometry: With few exceptions, there are adequate numbers of neoplastic cells in the PB for FCM and other special studies. FCM is, by far, the most important diagnostic technique for mature lymphocytic leukemias. With B-cell leukemias, FCM can verify clonality by demonstrating immunoglobulin (Ig) light chain restriction and can render an immunophenotypic profile that is diagnostically specific. With T-cell leukemias, FCM typically demonstrates an immunophenotype that is sufficiently abnormal to identify the process as neoplastic and can provide important clues to the specific diagnosis. NK-cell neoplasms express specific markers that may not be included in routine FCM panels and must be specifically requested, usually in the form of an "LGL panel."

Other special studies (FISH, PCR, Southern blot) are useful for demonstrating T-cell neoplasia by detecting clonal T-cell receptor (TCR) gene rearrangements. In difficult cases of B-cell leukemias, this can be accomplished using Ig gene rearrangement studies. These studies are not helpful with most NK-cell leukemias. An exception is the use of molecular studies to demonstrate a clonal pattern of X-chromosome inactivation in female patients with NK-cell neoplasms.

Chronic Lymphocytic Leukemia (CLL)

CLL is a neoplasm composed of small mature-appearing lymphocytes with small numbers of admixed prolymphocytes and accounts for 90% of chronic lymphoid leukemias in the United States and Europe. Small lymphocytic lymphoma (SLL) refers to a neoplasm with the same morphology and immunophenotype but affecting only solid tissues (non-leukemic disease). When insufficient information is available to make this distinction, the diagnosis CLL/SLL is appropriate. SLL accounts for 7% of non-Hodgkin lymphomas. The majority of CLL/SLL patients are 50 years of age or older, and the incidence increases with age. Men are affected twice as often as women.

Clinical: By definition, the PB and BM are involved in all cases. Other possible sites of involvement include the following:

- LNs
- Liver
- Spleen
- Skin
- Breast
- Ocular adnexa

Most patients are asymptomatic and discovered by routine laboratory work, but some have fatigue, weakness, weight loss, hepatosplenomegaly, lymphadenopathy or extranodal lesions. Fever, night sweats and recurrent infections may occur as the disease progresses. Some patients develop autoimmune hemolytic anemia or immune thrombocytopenia or neutropenia.

PB findings: The neoplastic lymphocytes are small and mature in appearance with clumped, ropy chromatin with no or indistinct nucleoli and a thin rim of clear or lightly basophilic cytoplasm. Small numbers of intermixed prolymphocytes are larger with loosely condensed chromatin, a single distinct nucleolus and moderate amounts of basophilic cytoplasm. Smudged and damaged cells, some vaguely resembling straw baskets (basket cells), are common. There is no strict criterion for the absolute lymphocyte count for the diagnosis of CLL, but it is usually approximately $5 \times 10^9/L$ or greater and may climb as high as $500 \times 10^9/L$ in advanced cases. Anemia (Hgb <11 g/dL), usually normochromic and normocytic, is seen in 15-20% of patients at the time of diagnosis, and thrombocytopenia (<100 $\times$ $10^9/L$) in approximately 10%. In patients with autoimmune hemolytic anemia, spherocytes or increased polychromatophilia may be observed.

BM findings: The appearance of the cells in the BM aspirate is similar to that in the PB but may be quite subtle and easily missed. The pattern of involvement of the BM biopsy may be interstitial, nodular, combined interstitial and nodular or diffuse replacement. Interstitial and nodular are more common in early disease, while the diffuse pattern is more often encountered in advanced cases. Proliferation centers, which are characteristic with SLL involvement of the LNs, are rarely seen in BM sections. Paratrabecular involvement is uncharacteristic. Atypical morphology usually cannot be appreciated on the biopsy.

Morphologic variants of CLL are referred to as "atypical CLL" and usually occur in one of two general patterns:

1. CLL with a mixture of small lymphocytes and large lymphocytes with condensed chromatin and moderate amounts of basophilic cytoplasm.
2. CLL with increased prolymphocytes (CLL/PL), >10% but <55% of lymphoid cells; this definition excludes cases of prolymphocytoid transformation.

Nuclear atypia may occur in some cases and must be differentiated from a leukemic phase of other lymphomas.

Transformation: Prolymphocytic transformation is an increase in the number of prolymphocytes (≥55% of lymphoid cells). Prolymphocytic transformation is extremely rare, and it is frequently accompanied by thrombocytopenia, lymphadenopathy, splenomegaly and resistance to therapy. Richter syndrome (or transformation) refers to the transformation of CLL to a high-grade lymphoma, most commonly diffuse large B-cell lymphoma (DLBCL), and occurs in about 5% of cases. Richter transformation to DLBCL is morphologically characterized by sheets of centroblastic, immunoblastic and/or paraimmunoblastic cells, which may also be seen in the PB. Rarely (0.5% of cases), CLL may be associated with Hodgkin lymphoma as a Richter transformation or otherwise, seen either as scattered Reed-Sternberg cells in a background of CLL or as discrete areas of Hodgkin lymphoma interspersed with CLL.

Other laboratory findings: The direct antiglobulin test (DAT) is positive in the 10-15% of patients who develop autoimmune hemolytic anemia, usually due to a warm-reacting IgG antibody. A positive DAT is also seen in approximately 10% of patients without evidence of autoimmune hemolytic anemia. Autoimmune thrombocytopenia and neutropenia may also be seen, but the responsible antibodies are difficult to detect. Serum protein electrophoresis (SPEP) demonstrates hypogammaglobulinemia of all Ig classes at some time in the disease course in almost all patients. About 5% of patients have a monoclonal protein, usually IgM.

Immunohistochemistry (IHC) plays a secondary role in the diagnosis of CLL. Immunohistochemistry for CD20 can be used to highlight an inapparent interstitial population of neoplastic lymphocytes in the BM biopsy. When the patient has been treated with rituximab, an anti-CD20 drug, this technique will usually not work. A good alternative is CD79a. IHC for FMC7 is usually negative in CLL and positive in most other B-cell lymphoproliferative disorders. Cytoplasmic Ig is only demonstrable in 5% of cases; therefore, IHC for κ and λ is a low-yield practice. Negative staining with bcl-1 (cyclin D1) is useful for differentiating CLL from MCL.

Flow cytometry is the most important special technique for the diagnosis of CLL. The immunophenotype of CLL is given in Table 9.2. Referring to Table 9.8 at the end of this chapter, it can be seen that FCM may be used to differentiate the mature B-cell leukemias from one another and from lymphomas with overlapping features. ZAP-70 is an intracellular tyrosine kinase involved in T-signaling that is

---

Table 9.2. Immunophenotype of CLL by flow cytometry

- Surface Ig dim or negative
- CD5 (+)
- CD19 (+)
- CD20 weak (+) (CD19 expression > CD20 expression is virtually pathognomonic)
- CD22 weak (+)
- CD79a (+)
- CD23 (+)
- CD11c weak (+)
- FMC7 (-)

Table 9.3. Recurrent chromosomal abnormalities in CLL

Genetic Abnormality	Comments
Trisomy 12	Present in 20%; atypical morphology; aggressive clinical course
Deletions at 13q14	Present in 50%; associated with long survival
Deletions at 11q22-23	Present in 20%; extensive lymphadenopathy; poor survival
Deletions at 6q21	Present in 5%
Deletions at 17p13	Present in 10%
Ig heavy chain variable region gene mutation	Associated with better prognosis
TP53 abnormalities	Associated with poor prognosis

a negative prognostic finding in CLL that can be demonstrated by FCM. CD38 positivity is also associated with a poor prognosis.

Genetics: Ig heavy and light chain genes are clonally rearranged. Chromosomal abnormalities are detected in 80% of cases by FISH and in 50-60% of cases by standard metaphase cytogenetics. A number of recurrent genetic lesions may be seen; most have prognostic significance (Table 9.3). Many reference laboratories offer FISH panels that include most of the important genetic abnormalities.

Prognosis: For practical purposes, CLL is an incurable disease that is usually indolent with prolonged survival. Treatment is withheld from most asymptomatic patients early in the disease course. In general, patients diagnosed when more than 60 years of age will die from unrelated causes and younger patients from the CLL, usually due to infection. The overall median survival is seven years. Several factors have an adverse influence on survival, including the following:

- CLL/PL variant
- Diffuse involvement of the BM
- Richter or prolymphocytic transformation
- CD38 positivity
- ZAP-70 positivity
- Chromosomal abnormalities (trisomy 12, deletions at 11q22-23, TP53 abnormalities)

The most important predictor of disease outcome is the stage. Two staging systems are commonly used, the Rai and Binet, which are depicted in Tables 9.4 and 9.5.

B-Cell Prolymphocytic Leukemia (B-PLL)

B-PLL is a are neoplasm (1% of lymphocytic leukemias) in which >55% of the lymphocytes in the PB are prolymphocytes, and usually >90%. Prolymphocytic transformation of CLL and variant CLL with increased prolymphocytes (CLL/PL) are excluded. The PB, BM and spleen are involved. Most patients are over 60 years of age, and men are affected almost twice as often as women.

Clinical: The typical patient presents with weakness, weight loss and abdominal discomfort. Physical examination reveals splenomegaly without lymphadenopathy. Some patients have a serum monoclonal protein.

Table 9.4. Rai staging system for CLL

Stage	Risk and Median Survival
Stage 0: lymphocytosis in PB and BM only	Low; 10 years
Stage I: lymphocytosis plus enlarged lymph nodes	Intermediate; 6 years
Stage II: lymphocytosis plus enlarged liver and/or spleen	Intermediate; 6 years
Stage III: lymphocytosis plus anemia (Hgb<11g/dL); lymph nodes, spleen or liver may be enlarged	High; 2 years
Stage IV: lymphocytosis and thrombocytopenia; anemia and organomegaly may be present	High; 2 years

From Rai KR. Clinical staging of chronic lymphocytic leukemia. Blood 1975 46:219-234.

PB findings: The absolute lymphocyte count is usually >100 x 10^9/L. The prolymphocytes are intermediate in size, that is, about twice the size of a normal lymphocyte, with a central prominent nucleolus and a small amount of basophilic cytoplasm. Sometimes the nucleus is indented. Anemia and thrombocytopenia are found in 50% of cases.

BM findings: B-PLL involves the BM in a diffuse or, sometimes, a mixed nodular and interstitial pattern.

IHC plays a secondary role in the diagnosis of B-PLL in most cases, but a B-cell stain such as CD20 can be helpful for identifying subtle interstitial involvement or estimating the amount of tumor involvement. In some cases, differentiation from blastic MCL may be quite difficult, both morphologically and by FCM. IHC for bcl-1 will be negative in B-PLL but demonstrate nuclear positivity in MCL.

Flow cytometry: B-PLL strongly expresses surface IgM (sIgM) and may be positive or negative for sIgD. Pan B-cell markers are also present, including CD19, CD20, CD22, CD79a, CD79b and FMC7. Positivity for CD79b and FMC7

Table 9.5. Binet staging system for CLL

Stage	Median Survival
A: Hgb ≥10 g/dL; plt ≥100 x 10^9/L; <3 anatomic sites* involved	>120 months
B: Hgb ≥10 g/dL; plt ≥100 x 10^9/L; ≥3 anatomic sites involved	61 months
C: Hgb <10 g/dL; plt <100 x 10^9/L	32 months

*Anatomic sites: inguinal lymph nodes, axillary lymph nodes, cervical lymph nodes, liver or spleen. From Binet JL. A new prognostic classification of chronic lymphocytic leukemia derived from a multivariate survival analysis. Cancer 1981 48:198-206.

are helpful for differentiating from prolymphocytic transformation of CLL not expressing these antigens. Approximately one-third of cases of B-PLL are CD5 positive, which complicates differentiation from MCL. B-PLL is typically negative for CD23, similar to MCL, but rare cases are positive. IHC (BM biopsy) or FISH (PB, BM aspirate, BM clot section) for bcl-1 (positive in MCL) resolves ambiguous FCM findings.

Genetics: Ig genes are clonally rearranged. Abnormalities of TP53 are present in about one-half of cases. The t(11;14)(q13;q32) abnormality has been reported to be positive in 20%. However, these cases most likely represent examples of blastic MCL and, in fact, detection of this abnormality by FISH or PCR of the PB or BM aspirate is a reliable way to prove the diagnosis of MCL in difficult cases. Deletions of 11q23 and 13q14 are identified in some cases by FISH.

Differentiation from blastic MCL is explained in a piecemeal fashion above but deserves summarization. The t(11;14) abnormality, also known as the bcl-1 gene or referred to by its protein product cyclin D1, can be demonstrated in the paraffin-embedded BM biopsy material by positive IHC staining in a nuclear pattern or, alternatively, by FISH of the PB, BM aspirate or BM clot section. Recall that FISH does not work on decalcified specimens. Bcl-1 positivity, for practical purposes, is specific for MCL. Regarding FCM, it should be kept in mind that both B-PLL and MCL are usually CD23 negative and that some cases of B-PLL are CD5 positive while some cases of MCL are CD5 negative. Therefore, it is prudent to assess bcl-1 in virtually all cases.

Prognosis: B-PLL is an aggressive disease that responds poorly to therapy and has short survival times.

Hairy Cell Leukemia (HCL)

HCL is a rare neoplasm, accounting for approximately 2% of lymphomas, which occurs in the middle-aged to elderly with a male to female ratio of 5:1. One of the most important aspects of HCL is that it is a treatable, usually curable, cause of cytopenias that is morphologically subtle and easily overlooked. In particular, on both clinical and morphological grounds, HCL can be mistaken for aplastic anemia, which has a poor prognosis and is sometimes treated with BM transplantation that carries the risk of numerous complications including death. Despite its rarity, HCL should be considered in the differential diagnosis of virtually all cytopenia evaluations.

Clinical: Most patients present with pancytopenia, which may be accompanied by opportunistic infections or vasculitis. Splenomegaly occurs in 80%. Monocytopenia is a characteristic finding.

PB findings: Hairy cells are one to two times the size of normal small lymphocytes and have oval or reniform nuclei with spongy chromatin and no or inconspicuous nucleoli. The cytoplasm is moderate to abundant and pale blue in color with indistinct borders or "hairy" projections. Hairy cells are present in the PB in nearly all cases but are difficult to find and are sometimes identified only on retrospective examination of the blood smear after identification of the disease in the BM. Classically, hairy cells are identified by positive staining for tartrate-resistant acid phosphatase (TRAP), but this technique has been largely replaced by immunophenotyping with FCM. CBC abnormalities seen with HCL are listed in Table 9.6.

Table 9.6. CBC abnormalities seen with HCL

Abnormality	Incidence
Pancytopenia	50%
One- or two-line cytopenia	25%
Neutropenia	75%
Monocytopenia	Nearly 100%
Anemia	Most

BM morphology: The BM aspirate is usually a "dry tap," due to reticulin fibrosis. In the BM biopsy, the pattern of involvement is interstitial or patchy and there is usually preservation of some of the fat and hematopoietic cells. The neoplastic cells are slightly larger than normal lymphocytes with oval or reniform nuclei and abundant cytoplasm with distinct cell borders imparting a "fried egg" appearance to the individual cells or a "honeycomb appearance" to groups of cells viewed on low power. Mitoses are scant. When the BM is involved by small amounts of tumor in an interstitial pattern, the neoplastic cells can blend imperceptibly with normal BM elements. IHC stains such as CD20 are invaluable in such situations. Reticulin fibrosis is common and may be seen using special stains such as Snook's reticulin stain. Advanced cases may show diffuse solid involvement of the BM. In 20% of cases the BM is markedly hypocellular, making differentiation from aplastic anemia quite difficult. A CD20 immunostain should be done on the BM biopsy specimen in all cases of suspected aplastic anemia. The decrease in the hematopoietic cells in such cases is theorized to be due to cytokines produced by the neoplastic lymphocytes.

Hairy cell variant: Rare cases of HCL demonstrate cells in the PB that resemble prolymphocytes, while the morphology in the BM and spleen remain typical. These patients present with leukocytosis, usually 50 x 10^9/L or greater, and lack monocytopenia. The variant neoplastic cells differ immunophenotypically from typical HCL cells in that they usually lack CD25 and sometimes lack CD103 expression. TRAP staining is often negative. The prognosis in these cases is somewhat poorer than that of typical HCL, and the patients do not respond to therapy with nucleoside analogues such as pentostatin or cladribine or to therapy with interferon-α2b.

Immunohistochemistry: As previously stated, IHC stains are extremely helpful in cases with low volume interstitial involvement of the BM and cases with marked hypocellularity. CD20 and DBA.44 are the most useful in such cases. HCL is negative for bcl-1.

Flow cytometry: Typical antigens expressed by HCL include CD25, CD11c, CD103 and FMC7. Common B-cell antigens such as CD19, CD20, CD22 and CD79a are also present. CD79b is usually negative. Pertinent antigens that are not expressed include CD5 (positive in CLL and MCL), CD10 (positive in follicular lymphoma) and CD23 (positive in CLL and absent in MCL). No one marker is specific for HCL.

Genetics: Ig heavy, light and variable region genes are clonally rearranged. There is no specific genetic abnormality in HCL.

Prognosis: The prognosis of HCL is good. HCL does not respond to conventional lymphoma therapy but, instead, is treated with special agents such as nucleoside analogues (pentostatin or cladribine) or interferon-α2b.

T-Cell Prolymphocytic Leukemia (T-PLL)

T-PLL is a rare (2% of small lymphocytic leukemias) neoplasm of cells that phenotypically resemble post-thymic T-lymphocytes. The median age at presentation is 70 years, and there is a slight male predominance. T-PLL involves the PB, BM, liver, spleen and skin.

Clinical: The most common presentation is that of hepatosplenomegaly with widespread lymphadenopathy. Skin infiltration is present in 20% of patients while a minority have a pleural effusion.

PB findings: The neoplastic prolymphocytes are medium to large with oval nuclei, moderately condensed chromatin and a prominent nucleolus. The cytoplasm is basophilic with occasional blebs and lacks granules. Some cells have irregular nuclei. Anemia, neutropenia and thrombocytopenia are common.

Morphologic variants: A small cell variant occurs in 25% of cases and is composed of small cells, as one might guess from the name, which lack nucleoli. The Sezary-like variant is rare and is composed of cells with cerebriform nuclei.

BM findings: The extent of BM involvement is variable and occurs in a mixed nodular and diffuse, or diffuse pattern. The cellular morphology in the aspirate is the same as that described for the PB.

Laboratory: The important laboratory results are negative findings. There is no monoclonal protein, the serum immunoglobulins are normal and serology for HTLV-1 is negative.

Cytochemistry: The neoplastic cells are strongly positive for alpha-naphthyl acetate esterase staining of the PB. Cytochemistry has been replaced by flow cytometry in most laboratories.

Immunohistochemistry: The IHC findings are nonspecific such as positivity for common T-cell antigens including CD3. T-PLL is negative for TdT which is helpful for the differentiation from precursor T-lymphoblastic leukemia/lymphoma. The usefulness of IHC is limited.

Flow cytometry: T-PLL is positive for CD2, CD3, CD5 and CD7 although the membranous CD3 may be weak and is negative in 20% of cases. Sixty percent of cases are CD4+/CD8−, 25% are CD4+/CD8+, a finding almost exclusive to T-PLL, and 15% are CD4−/CD8+. T-PLL is negative for TdT and CD1a.

Genetics: T-cell receptor (TCR) genes are clonally rearranged. The most common abnormality is inversion of chromosome 14 with a breakpoint in the long arm at q11 and q32, which is present in 80% of cases. Approximately 75% have abnormalities of chromosome 8 including idic(8p11), t(8;8)(p11-12;q12) or trisomy 8q. The t(11;14)(q11;q32) abnormality is seen in 10%. Deletions of 12p13 and 11q23 may be seen by FISH. Translocation (X;14)(q28;q11) is less common.

Prognosis: T-PLL has an aggressive course with a median survival of less than one year. Rare indolent cases have been reported.

T-Cell Large Granular Lymphocytic Leukemia (T-LGL)

T-LGL leukemia is a rare (2-3% of small lymphocytic leukemias) neoplasm in which there are PB LGLs, usually in the range of 2-20 x 10^9/L, for greater than six months duration without a known cause. In addition to the PB, the BM, liver and spleen are also involved. LN enlargement is rare. The age range of patients is broad: 4 to 88 years, with a median of 60 years. Cases with the natural killer cell (NK-cell) phenotype are excluded and placed with NK-cell disorders in the WHO classification scheme.

Clinical: Approximately 60% of patients are asymptomatic at the time of presentation and come to the attention of the clinical hematologist because of findings on routine laboratory testing. When clinical symptoms are present, the most common are related to infections due to neutropenia. The most common findings on physical examination are splenomegaly (seen in 35%) and hepatomegaly (seen in 20%). There is an association between T-LGL and rheumatoid arthritis. Hypergammaglobulinemia is a common finding.

PB findings: Neoplastic LGLs are similar to, if not indistinguishable from, reactive LGLs that are slightly enlarged lymphocytes with abundant cytoplasm containing fine to coarse azurophilic granules. Leukocytosis is present in about 75% of cases with an absolute lymphocyte count ranging from about 2 to 20 x 10^9/L. There is no strict criterion for an absolute lymphocyte count to make the diagnosis of T-LGL leukemia, but a good rule of thumb threshold is >5 x 10^9/L. Severe neutropenia with or without anemia is present in approximately 50% of patients and thrombocytopenia in about 20%. Red blood cell hypoplasia with severe anemia has been reported.

BM findings: The pattern of BM involvement is usually interstitial and is rarely nodular. A peculiar characteristic finding is the aligning of neoplastic cells along blood vessels that are nicely highlighted with a CD3 IHC stain and not seen with reactive LGLs. The amount of BM involvement is usually less than 50%, and the maturation sequence of the normal hematopoietic cells is fairly uniform. As previously mentioned, severe red cell hypoplasia has been reported.

IHC plays a secondary role in the evaluation of the BM in T-LGL leukemia. Stains for common T-cell markers such as CD3 can highlight an otherwise inapparent interstitial population of neoplastic cells. The previously described phenomenon of T cells aligned along blood vessels is a finding highly supportive of neoplasia that can generally only be appreciated with IHC.

Flow cytometry: The typical immunophenotype of T-LGL leukemia is CD3+, TCRαβ+ and CD4−/CD8+. Rare variants include the following:

- CD3+, TCRαβ+ and CD4+/CD8−
- CD3+, TCRαβ+ and CD4+/CD8+
- CD3+, TCRγδ+ and CD4−/CD8+ or CD4−/CD8−

Most cases are positive for CD57 and TIA-1 and variably express CD116 and CD11b.

Genetics: TCR genes are usually clonally rearranged. In most, TCR-β is rearranged while TCR-γ is rearranged in a minority. It is rare to see chromosomal abnormalities on standard metaphase cytogenetic studies.

Prognosis: T-LGL leukemia is usually an indolent disease. In fact, some think that it is a reactive increase in clonal T cells rather than a neoplasm. Neutropenia related to T-LGL leukemia causes morbidity but rarely mortality. An aggressive clinical course is seen occasionally, and a minority of cases undergo transformation to an aggressive T-cell lymphoma.

Aggressive NK-Cell Leukemia

This rare neoplasm of NK-cells typically involves the PB, BM, spleen and liver, although any organ may be affected. Aggressive NK-cell leukemia usually occurs in teenagers and young adults, and Asians more frequently than Caucasians.

Clinical: Most patients present with constitutional symptoms related to cytopenias. Hepatosplenomegaly is commonly found on physical examination, and lymphadenopathy is sometimes present. Patients may have a coagulopathy, hemophagocytic syndrome or multiorgan failure. The multiorgan failure is felt to be due to elevated serum FAS ligand. Some cases evolve from extranodal NK/T-cell lymphoma or indolent NK/T-cell lymphoproliferative disorders. There is a strong association between aggressive NK-cell leukemia and Epstein-Barr virus (EBV). Rarely, patients demonstrate hypersensitivity to mosquito bites.

PB findings: The leukemic cells can comprise from a few to more than 80% of the WBCs. The neoplastic NK-cells are slightly larger than LGLs and have nuclear irregularities, hyperchromasia, indistinct or prominent nucleoli, and abundant pale to deeply basophilic cytoplasm that contains fine or coarse azurophilic granules. Anemia, neutropenia and thrombocytopenia are common.

BM findings: The pattern of BM involvement may be massive, focal or subtle. Intermixed histiocytes and hemophagocytosis are sometimes present.

IHC does not play a primary role in identifying the neoplastic cells.

Flow cytometry: The leukemic NK-cells are positive for CD2, CD56, TIA-1 and granzyme B. They are negative for CD3 but positive for $CD3_\epsilon$. Some cases are CD116 and CD16 positive. CD57 is usually negative.

Genetics: TCR genes are germline. Clonality can be demonstrated by the pattern of X chromosome inactivation in female patients. EBV in clonal form is present in most cases. There are no specific cytogenetic findings, but abnormalities such as del(6)(q21;q25) may be seen.

Prognosis: As the name implies, aggressive NK-cell leukemia has an aggressive course that is fatal within one to two years or, sometimes, within days or weeks.

Indolent NK-Cell Lymphoproliferative Disorder

Synonyms for this disorder include chronic NK-cell lymphocytosis and NK-cell large granular lymphocytosis. It is debated whether this disorder is reactive or neoplastic. Most of the affected patients are adults.

Clinical: Patients are usually asymptomatic but rarely have vasculitis or nephrotic syndrome. They notably lack hepatosplenomegaly and lymphadenopathy. Unlike T-cell LGL leukemia, there is no association with RBC aplasia, neutropenia or rheumatoid arthritis.

Flow cytometry: The NK-cells in this disorder are positive for CD2, CD56, CD16 and CD57 and are negative for CD3, CD4, CD8 and EBV.

Genetics: The TCR genes are germline.

Prognosis: Indolent NK-cell lymphoproliferative disorder is usually non-progressive but rarely evolves to aggressive NK-cell leukemia. As previously stated, it is unclear whether this disorder is reactive or neoplastic.

Adult T-Cell Leukemia/Lymphoma (ATLL)

ATLL is caused by a retrovirus, human T-lymphotrophic virus, Type 1 (HTLV-1), which is endemic to Japan, the Caribbean basin and Central Africa with sporadic cases in the Southeastern United States. HTLV-1 is transmitted in breast milk or by exposure to blood or blood products. Most patients are infected early in life followed by a long latency period with a median age at disease presentation of 55 years. Not all infected patients develop ATLL; therefore, HTLV-1 alone is not sufficient to cause malignant transformation. Additional genetic alterations are required. The male to female ratio is 1.5:1.

Clinical: ATLL patients usually present with diffuse lymphadenopathy and involvement of the PB. The extent of the PB involvement does not correlate with the extent of BM involvement; therefore, sites other than the BM supply neoplastic cells to the PB. Extralymphatic involvement is common, especially of the skin (50%) but also of the lung, liver, gastrointestinal tract and central nervous system. Lytic bone lesions and hypercalcemia may also be present. Four clinical variants are recognized:

- Acute variant: Leukemic phase with markedly elevated WBC count (20-50 x 10⁹/L), skin rash, lymphadenopathy and hypercalcemia with or without lytic bone lesions, hepatosplenomegaly, constitutional symptoms, increased LDH or eosinophilia. T-cell immunodeficiency occurs with opportunistic infections such as *Pneumocystis carinii* and strongyloides.
- Lymphomatous variant: Characterized by lymphadenopathy. Similar to acute variant, but no PB involvement and hypercalcemia less common.
- Chronic variant: Skin lesions, usually an exfolliative rash. Hypercalcemia is absent. Atypical lymphocytes are rare in the PB.
- Smoldering variant: WBC is normal with <5% circulating neoplastic lymphocytes. Skin and pulmonary lesions are common. Hypercalcemia is absent.
- In 25% of cases there is evolution from chronic to smoldering to the acute variant over a long period of time.

PB findings: Neoplastic lymphocytes are present in the PB in the acute and lymphomatous variants. The malignant cells are medium to large with marked nuclear irregularities, sometimes termed "flower cells", with coarse and clumped nuclear chromatin and prominent nucleoli. The cytoplasm is scant, deeply basophilic and non-granular. In all cases, there are smaller numbers of blast-like cells with dispersed chromatin. There are sometimes giant cells with highly irregular nuclei that are conspicuous enough to be seen on scanning power. Rare cases have small atypical cells with pleomorphic nuclei. The clinical course does not correlate with cell size.

BM findings: The BM usually demonstrates patchy, sparse to moderate involvement. When there are no neoplastic cells found in the BM, osteoclastic activity may be pronounced.

Immunohistochemistry: The most useful IHC stain in ATLL is CD3. The large cells may be positive for CD30 but are negative for ALK, TIA-1 and granzyme B.

Flow cytometry: Most cases are CD4+/CD8−, but rare cases are CD4−/CD8+ or CD4+/CD8+. Almost all are strongly positive for CD25.

Genetics: The neoplastic cells possess clonally integrated HTLV-1. TCR genes are clonally rearranged.

Prognosis: Survival in the acute and lymphomatous variants ranges from two weeks to over one year, with most patients succumbing to infection (*Pneumocytis carinii* pneumonia, Cryptococcus meningitis, herpes) and hypercalcemia. The chronic and smoldering variants have a protracted course but have the potential to progress to the acute or lymphomatous variants.

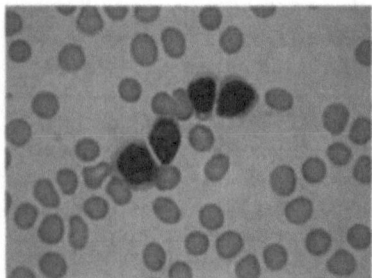

Figure 9.1. CLL, PB with increased mature-appearing lymphocytes (500x).

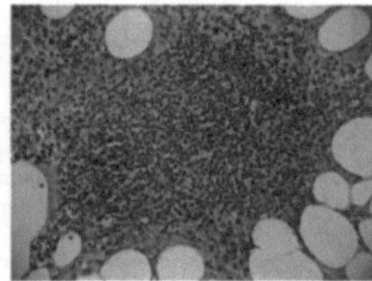

Figure 9.2. CLL, BM biopsy with intertrabecular aggregate of small lymphocytes (100x).

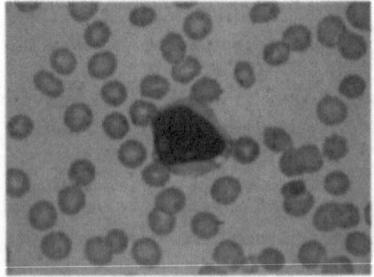

Figure 9.3. B-PLL, PB (500x).

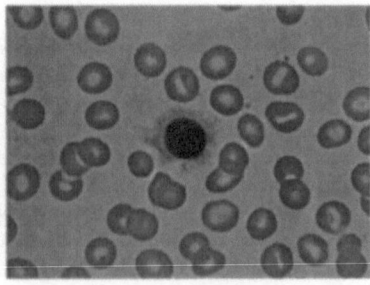

Figure 9.4. HCL, PB (500x).

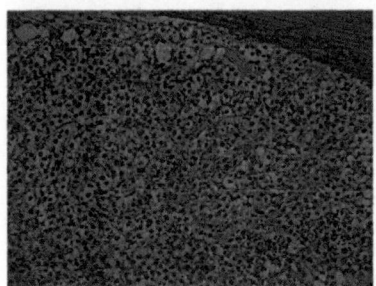

Figure 9.5. HCL, BM biopsy (200x).

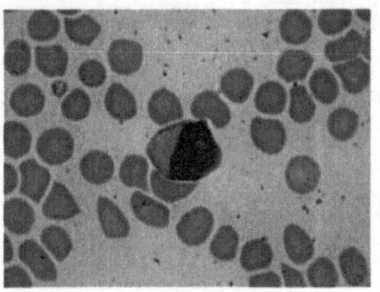

Figure 9.6. T-PLL, PB (500x).

9

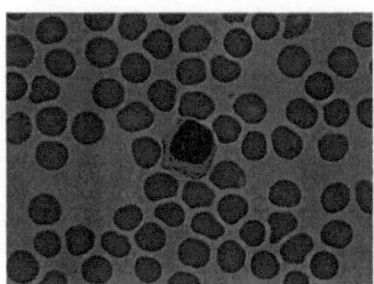

Figure 9.7. T-LGL, PB (500x).

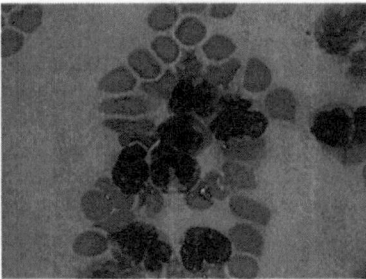

Figure 9.8. Adult T-cell leukemia/lymphoma (PB, 500x).

CHAPTER 10

Plasma Cell Disorders

Introduction

The presence of clonal plasma cells (PCs) can be associated with a variety of manifestations, ranging from the asymptomatic laboratory finding of mono-clonal gammopathy of undetermined significance (MGUS) to the aggressive fatal course of plasma cell leukemia (PCL). Bone marrow (BM) plasmacytosis may also be seen as a reactive phenomenon with clinical findings that are dominated by the underlying process. Lymphoplasmacytic lymphoma (LPL) will also be discussed in this chapter, because it has in common with the other PC disorders the production of monoclonal immunoglobulin as well as some morphologic features.

Abnormal Protein Production

Theoretically, any B-cell neoplasm can secrete immunoglobulin (Ig). Being the end-stage of B-cell differentiation, PCs secrete Ig as part of their normal function. Plasma cell neoplasms usually secrete abnormal Igs which are important to the diagnosis of these disorders and which are responsible for many of their clinical manifestations. Abnormalities of Ig production include:

- Production of a single class of Ig (G, A, D, E or M) with light chain restriction (κ or λ) at the expense of decreased production of normal Igs
- Production of incomplete Ig molecules such as light chains only
- The absence of Ig production
- Failure to secrete the Igs produced
- Production of Igs with mutational or deletional structural defects

Multiple synonyms are used for these proteins, including monoclonal gam-mopathy, dysproteinemia, paraproteinemia, M-component and M-protein. Do not confuse the "M" in M-protein, which stands for monoclonal, with IgM. Light chains only are known as Bence Jones proteins.

Laboratory testing for monoclonal proteins: The detection of these proteins is usually accomplished by protein electrophoresis (PEP or EP) of either the serum (SPEP) or urine (UPEP). Full-size, intact M-proteins appear as a dense spike on electrophoresis (M-spike), usually in the γ-globulin area, occasionally in the β-globulin area and, rarely, in the α_2-globulin area. The normal polyclonal Igs are usually decreased. When the M-protein is not present in large quantities, as is common with IgD, IgE and light chain only gammopathies, they may not be detected by SPEP. The only finding on SPEP may be hypogammaglobulinemia.

Bone Marrow: A Practical Manual, by Daniel A. Cherry and Tomislav M. Jelic.
©2011 Landes Bioscience.

Bence Jones proteins are about one-fourth the size of intact Ig molecules and are able to pass through the glomerular filter, resulting in rapid clearing from the serum. For this reason, 24-hour urine collection with EP should be done in all suspected cases.

Once an M-protein is identified, it should be characterized by more refined methods of electrophoresis, most commonly immunofixation electrophoresis (IFE). Laboratories that deal with these patients should establish a protocol for reflexively performing IFE on all specimens in which an M-protein is identified.

Quantitative values of M-protein in 24-hour urine collections are of prognostic value and are used for this purpose and/or surveillance.

Identification of PC clonality: The qualities of these abnormal proteins, such as being of a single Ig class and light chain restriction, inform on the clonality of the PCs that produce them. Monoclonality can also be easily demonstrated by immunohistochemistry (IHC) for cytoplasmic light chain restriction or in a more cumbersome way by the special flow cytometry technique of permeabilization of cytoplasmic membrane to allow fluorochrome-labeled antibodies to reach intracytoplasmic immunoglobulins. The normal κ to λ light chain ratio is approximately 2:1. By IHC, κ light chain monotypism is said to be present when the κ:λ is 5:1 and λ monotypism when the κ:λ is 1:2. Exact quantitation of this ratio is difficult; however, on a practical level the predominance of either κ or λ is usually striking.

Peripheral blood (PB) evidence of dysproteinemia may be seen in the form of rouleaux, which is the sticking together of RBCs in a pattern that resembles stacked coins. Rarely, the M-protein is seen as a faint purple background on Wright stained blood smear slides.

Consequences of M-protein: As previously referenced, M-proteins are responsible for many of the clinical manifestations of these disorders. The deposition of M-protein in various tissues can lead to dysfunction and/or organomegaly. High levels of paraprotein in the PB can cause hyperviscosity. The binding of paraprotein to coagulation factors can result in bleeding. Specific clinical consequences of these proteins are discussed more fully with the individual disorders.

Bone marrow evaluation in PC disorders: The most common reasons for BM sampling in which PC disorders are discovered include:

- To work up a patient with an M-protein
- To rule out myeloma in a patient with lytic bone lesions
- To evaluate cytopenia(s) in patients in which a PC disorder is not specifically suspected

Often multiple indications of a PC disorder are known at the time of the BM study, such as an M-protein discovered in the setting of lytic bone lesions.

Plasma Cell Myeloma (PCM)

PCM, sometimes referred to as multiple myeloma because of the multicentricity of the lesions, is an aggressive malignant neoplasm of PCs. Plasma cell myeloma is, by far, the most common PC neoplasm and is the most common lymphoid malignancy in blacks and the second most in whites. Blacks are affected more commonly than whites, there is a slight male predominance and the median age at diagnosis is 69 years.

Clinical findings: The majority of the clinical findings with PCM are related to skeletal involvement which occurs most commonly in the bones with the most active

---

Table 10.1. Clinical findings related to skeletal involvement by PCM

- Lytic lesions
- Pathologic fractures
- Generalized osteoporosis
- Neurological symptoms related to spinal involvement
- Hypercalcemia
- Shortening of stature due to vertebral collapse
- Cytopenias related to BM replacement

10

marrow, including the vertebrae, ribs, skull, pelvis, femur, clavicle and scapula. Bone pain is the most common presenting clinical symptom, usually of the lower back. About 75% of patients have skeletal abnormalities on X-ray. A summary of skeletal findings is given in Table 10.1. Patients may also present with renal failure due to the deposition of M-protein in the kidneys or symptoms related to cytopenias, such as fatigue secondary to anemia. Infections are common because of the suppression of the production of normal Igs. Physical findings are usually vague, but some patients have organomegaly due to plasmacytoma or amyloidosis, or bleeding caused by the binding of M-proteins with coagulation factors. Occasionally patients are incidentally discovered to have monoclonal gammopathy on laboratory work done for unrelated reasons. Skin lesions composed of plasma cell infiltrates occur very rarely.

Monoclonal proteins are present in 99% of cases with myeloma. In addition to SPEP, an UPEP should be done on a 24-hour urine collection in all cases to evaluate for Bence Jones proteins (light chain only disease). Table 10.2 lists the various abnormal proteins seen with plasma cell myeloma and their frequencies. Note that some cases have two clonal proteins (biclonal), only light chain (Bence Jones protein) production or are non-secretory. Light chains may also be produced in addition to intact M-protein. Kappa is more common than λ for all Ig classes except IgD. Of the non-secretory cases, some produce a protein that is detectable in plasma cells by flow cytometry or immunohistochemistry but fail to secrete it and, therefore, is undetectable by SPEP, UPEP or IFE. Other cases produce no Ig, resulting in plasma cells identified with CD138 by IHC or CD38 by FC that lack Ig or light chain expression.

Table 10.2. Monoclonal proteins with plasma cell myeloma

Protein	Frequency
IgG	50%
IgA	20%
Light chain only	15%
IgD	2%
Biclonal	1%
Non-secretory	1%
IgE	Rare
IgM	Rare

Table 10.3. Laboratory findings with MM

Laboratory Abnormality	Frequency
Anemia	60% at diagnosis
Thrombocytopenia	<20% at diagnosis worsening with disease progression
Leukocytosis	Rare
Thrombocytosis	Rare
Hypercalcemia	20% at diagnosis
Increased creatinine	35% at diagnosis
Hyperuricemia	>50% at diagnosis
Hypoalbuminemia	Present in advanced disease

Laboratory findings: In addition to monoclonal proteins and hypercalcemia related to skeletal involvement, a number of laboratory abnormalities may be present (see Table 10.3).

Peripheral blood findings: The most common and striking finding in the PB is rouleaux. Rarely, a faint purple background may be found when the M-protein is extremely high. PCs are found in the PB in 15% of patients and may demonstrate a variety of morphological appearances ranging from normal-appearing PCs, to lymphocyte-like cells to blast-like cells. Nucleated red blood cells or leukoerythroblastosis may be present. Plasma cell leukemia will be discussed subsequently.

Bone marrow aspirate: The BM aspirate and biopsy may be independently diagnostic. It is generally felt that BM sampling is necessary for the diagnosis of PCM, although an exception might be PCL. The PCs seen in the BM aspirate are mature and appear normal in the majority of cases with minimal atypia, including occasional multinucleated cells (defined as >2 nuclei since normal plasma cells may be binucleated) or mild nuclear membrane irregularities. Remember that normal PCs have eccentrically placed round nuclei with chromatin clumping in a clock-face arrangement, abundant basophilic cytoplasm and a perinuclear cytoplasmic hoff. A variety of cytoplasmic inclusions may be seen that are caused by the retention of Ig with distention of the endoplasmic reticulum (Table 10.4). Rod-like inclusions may also be seen. Nuclear pseudoinclusions, called Dutcher bodies, may also be seen and are due to cytoplasmic invaginations.

Plasmablastic morphology: These cases are defined by dispersed nuclear chromatin, high nuclear to cytoplasmic ratios, multinucleation and nuclear atypia. The median survival of patients with plasmablastic morphology is 10 months, compared to 35 months with conventional morphology. Sometimes the PCs are morphologically classified into mature, intermediate, immature and plasmablastic types; however, plasmablastic is the only type of prognostic significance. Neither immature nor plasmablastic morphology is seen with reactive plasmacytosis.

Bone marrow biopsy: in the BM biopsy specimen, PCs are seen in large (>6 cells) clusters in locations where they are not usually found (normal location is around arterioles) and displace normal marrow elements. Percentage involvement can be roughly estimated on the biopsy and is aided by IHC for CD138. Greater

Table 10.4. Cytoplasmic changes seen in plasma cells with PCM

Cytoplasmic Change	Morphologic Term
Pale, clustered grape-like inclusions	Mott or Morula cells
Round cherry-red inclusions	Russell bodies
Vermilion red staining in IgA myeloma	Flame cells
Overstuffed fibrils	Gaucher-like cells

than 30% PCs is strong evidence for a neoplastic rather than reactive process. Percentage involvement is prognostically important, as it reflects tumor burden. Light chain monotypism can be demonstrated with immunostaining (see abnormal protein production above). In cases with plasmablastic morphology, the malignant plasma cells may be difficult to recognize in the biopsy. In these instances and when there is lymphoid morphology, Dutcher bodies are an important clue to cell type. Reticulin fibrosis is present in about 9% of cases and is more common in light chain only disease. Coarse fibrosis is associated with diffuse marrow involvement and an aggressive clinical course.

Flow cytometry: In most cases, the neoplastic PCs are positive for CD38, CD138 and CD79a and most, but not all, lack CD19 and CD20. Normal plasma cells are positive for CD19 and negative for CD56/58, in contrast to neoplastic PCs that are negative for CD19 and positive for CD56/58. Occasional cases are positive for CD10 and/or co-express myelomonocytic markers. FCM typically detects surface antigens, but special techniques can be used that cause permeabilization of the cell membrane enabling the fluorochrome labeled antibodies to kappa and lambda to react with intracytoplasmic immunoglobulins and, thus, demonstrate monoclonality. The absence of surface Ig is evidence of malignancy but is nonspecific.

Genetics: Standard cytogenetics is problematic in myeloma because this technique relies on the presence of cells in metaphase. Because PCs are at the end point of B-cell differentiation, the proliferative rate is low and metaphase cells are rare. About 18 to 30% of cases have chromosomal abnormalities by cytogenetics which are associated with a poorer prognosis overall than the same lesions when identified by interphase techniques such as fluorescent in situ hybridization (FISH), because they infer a higher proliferative rate among the abnormal cells. By FISH nearly all cases have chromosomal abnormalities, including deletions, aneuploidy and translocations.

Ploidy: Nonhyperdiploidy is associated with more aggressive disease except for cases with t(11;14)(q13;q32). Nonhyperdiploidy is strongly associated with deletions of chromosome 13 and Ig heavy chain translocations (chromosome 14). Hyperploidy is a favorable prognostic indicator and is associated with trisomies of chromosomes 3, 5, 6, 7, 9, 11, 15, 17, 19 and 21 by cytogenetics and FISH.

Chromosome 13 deletions: Monoallelic loss of chromosome 13 (Δ13) or its long arm (Δ13q) is associated with a poor prognosis, which is of greater significance when detected by metaphase cytogenetics (M-Δ13q) as opposed to FISH (F-Δ13q).

Table 10.5. WHO criteria for PCM

- M-protein in serum or urine
- BM clonal PCs or plasmacytoma
- Related end organ tissue impairment (CRAB*)

*Calcium >10.5 mg/dL; **R**enal insufficiency (creatinine >2 mg/dL); **A**nemia (Hgb <10 g/dL); **B**one lesions. Adapted with permission from: Swerdlow SH, Campo E, Harris NL et al. World Health Organization Classification of Tumours of Haematopoietic and Lymphoid Tissues. Lyon: IARC, 2008:202, Table 10.05.

Ig heavy chain translocations: The Ig heavy chain gene is located on chromosome 14, and translocations involving this chromosome are seen in about 65% of cases of PCM. The most common is t(11;14)(q13;q32), which occurs in about 20% and which along with t(6;14)(p21;q32) is a favorable prognostic finding. The t(4;14)(p16.3;q32) and t(14;16)(q32;q23) abnormalities occur in approximately 15% and 5% of cases, respectively, and are associated with poor disease outcome. These translocations are usually only detected by FISH.

Deletions of 17p13 (p53): Deletions of p53 (locus 17p13) are present in about 10% of newly diagnosed cases of PCM and occur in increasing frequency with disease progression. These abnormalities are associated with a poor prognosis and are generally only detected by FISH.

A summary of the prognostic significance of genetic abnormalities is given in Table 10.8 in the section on prognosis below. It should be kept in mind that these genetic findings may be of little practical significance for the treatment and education of the patient in a community hospital setting.

Approach to the diagnosis of PCM: It can be seen from the WHO criteria (Table 10.5) that the diagnosis of PCM requires more than BM findings including chemistry and/or radiographic data. These data are usually readily available. Most hematopathologists will diagnose PCM if there are >30% PCs in the BM and these PCs have been demonstrated to be monoclonal by either IHC or FCM.

The principal tasks for the pathologist are simple and straightforward:

- Identify an increase in the number of PCs in the BM.
- Determine whether or not the PCs are monoclonal.
- Provide the morphological data necessary for the clinician to stage the disease (percentage involvement of BM biopsy).
- Assure that the genetic studies appropriate to the treatment setting are obtained.

Quantification of PCs: The number of PCs present in the BM can be determined on either the aspirate or biopsy. The percentage of PCs used to apply the WHO criteria is usually that obtained in the differential cell count of the aspirate, while the number used for staging of tumor burden (Table 10.6) is usually estimated on the biopsy. At times these numbers are discordant. Nodularity and focality may cause fewer PCs to be aspirated than is evident on the biopsy specimen. An interstitial distribution of the PCs may cause the deception of fewer PCs in the biopsy than seen in the aspirate, in which case a CD138 IHC stain is invaluable.

10

Table 10.6. Staging of tumor burden by percentage involvement of BM biopsy

Stage	% Involvement
I	<20%
II	20-50%
III	>50%

Regardless of tradition, the larger of the two numbers, aspirate or biopsy, should be used for both the application of the WHO criteria and staging.

Differentiating PCs from other cell types: Occasionally, an increased lymphoid cell population is evident and there is a question of LPL, chronic lymphocytic leukemia, marginal zone lymphoma or follicular lymphoma with plasmacytic morphology. The differentiation of these neoplasms from PCM can be accomplished by a variety of means. Morphologically, other lymphoid neoplasms usually demonstrate a mixture of plasmacytoid lymphocytes, mature-appearing lymphocytes and PCs as opposed to the relatively homogeneous morphology of myeloma. Immunophenotypically, plasmacytoid lymphocytic neoplasms almost always express CD19 and/or CD20 typically lacking with myeloma. Clinically, lymphadenopathy and/or splenomegaly, common findings with plasmacytoid lymphomas, are rare with myeloma and the skeletal symptoms common with myeloma are rare in plasmacytoid lymphomas. Chemically, elevated creatinine and calcium are not typical features of plasmacytoid lymphomas and IgM gammopathy, a defining feature of LPL, is extremely rare in myeloma.

Differential diagnosis of myeloma with plasmablastic morphology: Myeloma with plasmablastic morphology may pose a diagnostic problem in the biopsy specimen, but the PC differentiation is usually evident in the aspirate. If an adequate aspirate specimen is not available, then the differentiation of these cells from high-grade lymphoma or neoplastic myeloid cells can be done using immunophenotypic and clinical features. Neoplastic myeloid cells are positive for myeloperoxidase by IHC, and myeloid neoplasms are not usually associated with skeletal symptoms. The differentiation from high-grade lymphoma is accomplished in essentially the same manner as the differentiation from plasmacytoid lymphomas outlined above.

Demonstration of monoclonality: The differentiation of clonal (neoplastic) from polyclonal (reactive) PCs is one of the easier tasks in hematopathology. Clonal PCs produce Ig of a single class (G, A, D, E, M) with a single type of light chain (κ or λ). This Ig homogeneity can be identified by either IHC or FCM, as previously described.

Staging of tumor burden by morphology: Tumor burden as determined by the percentage involvement of the BM biopsy is prognostically significant and is used by some clinicians to follow response to therapy. This staging system is depicted in Table 10.6. An IHC stain for CD138 is extremely useful when making this estimation.

Prognosis: A variety of important prognostic factors have been identified for PCM (Table 10.7). The plasma cell labeling index (PCLI) is an important reproducible prognostic factor that is determined using a slide-based immunofluorescent assay but is seldom ordered by pathologists based in community hospitals.

Table 10.7. Prognostic factors for PCM

- Plasma cell labeling index (PCLI)
- β_2-microglobulin (β_2-M)
- Chromosomal abnormalities (see Table 10.8)
- C-reactive protein (CRP)
- Lactate dehydrogenase (LDH)
- Plasmablastic morphology
- Increased plasma cells in PB (3×10^6/mm^3)
- Increased BM angiogenesis
- Patient age
- Hemoglobin level
- Disease burden in BM
- Number of CD19$^+$ or CD4$^+$ cells in PB
- Presence of amyloidosis

10

Therefore, it is important to know ahead of time if the treating physicians desire this data. β_2-microglobulin correlates with tumor burden. Chromosomal abnormalities have been previously discussed. The prognostically important genetic lesions and their significance are given in Table 10.8. As previously stated, unfavorable chromosomal abnormalities identified by standard metaphase cytogenetics are more significant than the same abnormalities found by FISH because of the implied greater proliferative index. C-reactive protein (CRP) is a surrogate marker for interleukin-6 (IL-6), which is a growth factor for PCs and the elevation of which is associated with a poor prognosis. In contrast to IL-6 measurement, CRP levels are inexpensive and available in most laboratories. Elevation of lactate dehydrogenase (LDH) is a poor prognostic sign that is seen in <10% of myeloma patients. High-grade angiogenesis, as determined by increased vascularity identified with a Factor VIII immunostain, is associated with an unfavorable outcome but is not routinely evaluated.

Table 10.8. Prognostically important genetic lesions in PCM

Genetic Abnormality	Affect on Disease Outcome
Non-hyperdiploidy except that associated with t(11;14)(q13;q32)	Poor
Hyperdiploidy	Favorable
Chromosome 13 deletions	Poor
t(11;14)(q13;q32)	Favorable
t(6;14)(p21;q32)	Favorable
t(4;14)(p16.3;q32)	Poor
t(14;16)(q32;q23)	Poor
p53 deletions	Poor

Table 10.9. PCM staging system

Stage I
- Low M-component: IgG <5 g/dl, IgA <3 g/dl, urine Bence-Jones protein <4 g/dl
- Absent or solitary bone lesions
- Normal Hgb, serum calcium, and Ig levels (non-M-component)
- Median survival >60 months

Stage III: Any one or more of the following
- High M-component: IgG >7 g/dl, IgA >5 g/dl, urine Bence-Jones protein >12 g/dl.
- Advanced, multiple lytic bone lesions
- Hgb <8 g/dl, serum calcium >12 mg/dl
- Median survival 23 months

Stage II: Overall values between Stages I and III; median survival 41 months

Adapted with permission from: Swerdlow SH, Campo E, Harris NL et al. World Health Organization Classification of Tumours of Haematopoietic and Lymphoid Tissues. Lyon: IARC, 2008:207, Table 10.07.

Clinical staging is usually done by the clinician. The most common staging scheme used is a modification of the Durie-Salmon system, which is given along with the median survivals in Table 10.9.

Variants of Myeloma

Non-secretory myeloma: This rare (1%) myeloma variant was mentioned briefly with the description of abnormal protein production. In non-secretory myeloma, M-protein is produced but not secreted or, more rarely, no Ig is produced. In the former instance, monoclonal Ig can be identified by IHC or FCM but is not detected in either the serum or urine. In the latter instance, PCs as identified with CD138 by IHC or CD38 by flow cytometry, do not demonstrate Ig, which is a feature suggestive of neoplasia. In either case, the diagnosis will be missed if the patient is screened with only SPEP and UPEP and BM sampling is not done. These patients usually have a smaller number of PCs in the BM and less suppression of normal Ig production when compared to typical myeloma. Clinical features related to the deposition of abnormal Ig such as renal failure and hyperviscosity are absent, and infections are fewer.

Smoldering myeloma: In smoldering myeloma an M-protein is present at myeloma levels and the BM plasmacytosis is 10-30%, but the patients are asymptomatic and lack lytic bone lesions or other findings of myeloma (cytopenias, renal insufficiency and hypercalcemia). In some cases there is a small M-protein in the urine and normal Ig production is decreased. There may be many years between the diagnosis of smoldering myeloma and the development of overt myeloma. By current standards these patients are not treated until progression occurs. The WHO criteria for smoldering myeloma are given in Table 10.10.

Table 10.10. WHO criteria for smoldering myeloma

- Serum M-component at myeloma levels (>30 g/dL)
AND/OR
- 10% or more clonal PCs in BM
- No related organ or tissue impairment (CRAB: hypercalcemia, renal insufficiency, anemia, bone lesions*) or myeloma-related symptoms

Adapted with permission from: Swerdlow SH, Campo E, Harris NL et al. World Health Organization Classification of Tumours of Haematopoietic and Lymphoid Tissues. Lyon: IARC, 2008:202, Table 10.05.

Indolent myeloma: This variant is similar to smoldering myeloma except there may be up to three lytic bone lesions without fracture or bone pain. These patients are usually not treated unless progression occurs. The WHO criteria for indolent myeloma are given in Table 10.11.

Plasma cell leukemia: In this rare (2%) variant of myeloma, PB plasma cells are >2 x 10^9/L or 20% on the differential cell count. This finding may be present at the time of diagnosis or may evolve as a terminal stage of myeloma and tends to involve slightly younger patients. PCL is frequently associated with light chain only, IgD and IgE and less frequently with IgG or IgA. These patients may have many or all of the symptoms of typical myeloma, but osteolytic lesions and bone pain are less frequent and lymphadenopathy and organomegaly are more common. Renal failure is common. PCL has an aggressive course with short survival.

Monoclonal Gammopathy of Undetermined Significance (MGUS)

With MGUS an M-protein is present, but there is no evidence of myeloma, amyloidosis or Waldenstrom's macroglobulinemia. The incidence of MGUS increases with age, being approximately 1% in persons over 50 years and 3% in those older than 70 years. MGUS is a precursor lesion but not an obligate precursor lesion, in that 25% of patients develop a lymphoproliferative neoplasm such as myeloma, amyloidosis, or Waldenstrom's. The M-protein may be IgG (75%), IgM (15%) or IgA (10%), and the risk of neoplastic transformation is not dependent on the type of protein. The median interval to progression is 10 years with a risk of approximately 1% per year. These patients must be followed indefinitely. The WHO criteria for MGUS are given in Table 10.12.

Table 10.11. WHO criteria for indolent myeloma

Same as myeloma except:
- M-component: IgG <7 g/dl, IgA <5 g/dl
- ≤3 lytic bone lesions without compression fractures
- Normal Hgb, serum calcium, and creatinine
- No infections

Adapted with permission from: Jaffe ES, Harris NL, Stein H, Vardiman JW. WHO Classification of Tumours, Pathology and Genetics of Tumours of Haematopoietic and Lymphoid Tissues. Lyon: IARC, 2001:144, Table 6.08.

Table 10.12. WHO criteria for MGUS

- M-component present in serum <30 g/L
- BM clonal PCs <10% and low level of PC infiltration in trephine biopsy
- No related organ or tissue impairment (CRAB: hypercalcemia, renal insufficiency, anemia, bone lesions*) or myeloma-related symptoms

Adapted with permission from: Swerdlow SH, Campo E, Harris NL et al. World Health Organization Classification of Tumours of Haematopoietic and Lymphoid Tissues. Lyon: IARC, 2008:201, Table 10.04.

Primary Amyloidosis

In primary amyloidosis the secretion of Ig results in the deposition of β-pleated sheet substance in various body organs, resulting in organomegaly and sometimes organ dysfunction. A characteristic feature is apple-green birefringence seen on polarization when tissue sections are stained with Congo red. About 80% of patients have monoclonal Ig, and about 20% have overt myeloma. Approximately 15% of patients with myeloma will develop primary amyloidosis. The presence of amyloidosis worsens the prognosis of myeloma. Four types of amyloid may be deposited in tissues:

1. Primary or Ig light chain (AL) amyloidosis (myeloma-associated)
2. Secondary (AA) amyloidosis (inflammation-associated)
3. Familial (AF) amyloidosis
4. β_2-microglobulin (β_2M) amyloidosis (dialysis-associated)

This discussion refers to the first (AL) type although there are many generalities, such as the affect on body organs, which apply to all.

Organ involvement by amyloidosis: Amyloid deposition may occur in a variety of soft tissues and in bone. Sites commonly used for the biopsy diagnosis of amyloidosis include the BM, subcutaneous abdominal fat pad and rectum. Sites of involvement and associated clinical findings are depicted in Table 10.13. Bleeding may also occur as a result of binding of abnormal protein to coagulation Factor X.

Morphology: On hematoxylin and eosin staining, amyloid is seen as a waxy amorphous substance with cracking artifact in blood vessels, adipose tissue and BM interstitium. Foreign body reaction may occur around the deposits, and there may

Table 10.13. Organ involvement by amyloidosis

Site	Clinical Finding
Heart	Congestive heart failure
Liver	Hepatomegaly
Kidneys	Nephrotic syndrome and/or renal failure
Gastrointestinal tract	Malabsorption
Tongue	Macroglossia
Nerves	Sensorimotor peripheral neuropathy; loss of sphincter control

be increased numbers of PCs especially in the BM. The associated PCs demonstrate κ or λ monotypism. There is a characteristic appearance on staining with Congo red as described above.

Monoclonal Light and Heavy Chain Deposition Disease

These are rare disorders of adults that occur in association with myeloma or MGUS in which neoplastic PCs secrete abnormal Ig light or, less often, heavy chains that have undergone deletional or mutational structural alterations that promote their deposition in organs. Commonly involved organs include the kidneys, liver, heart, nerves, blood vessels and joints. This protein deposition causes organ dysfunction. The abnormal proteins may sometimes be seen on hematoxylin and eosin stained tissue sections as a refractile eosinophilic substance on basement membranes or in blood vessel walls but do not stain with Congo red. They can be identified by electron microscopy or immunofluorescence on kidney biopsies. PCs with κ or λ monotypism may be seen in the BM in accordance with the presence of myeloma or MGUS. The prognosis is very poor, with survival of only one or two years.

Osteosclerotic Myeloma

In this neoplasm of adults with a slight male predominance, there is BM plasmacytosis accompanied by thickening of the bony trabeculae. These patients may also have lymph node changes resembling the plasma cell variant of Castleman's disease. Osteosclerotic myeloma may occur as a component of POEMS syndrome:

    P: Polyneuropathy (sensorimotor demyelination)
    O: Organomegaly (hepatic, splenic)
    E: Endocrinopathy (diabetes, gynecomastia, testicular atrophy, impotence)
    M: Monoclonal gammopathy
    S: Skin changes (hyperpigmentation, hypertrichosis)

Clinical findings: Patients may present with a variety of symptoms usually related to polyneuropathy and endocrine dysfunction. Thrombocytosis is common, erythrocytosis may occur and anemia is rare.

Morphology: The BM demonstrates single or multiple lesions comprised of thickened trabecular bone with paratrabecular fibrosis containing entrapped PCs. The PCs contain monoclonal Ig of IgG or IgA heavy chain type. The light chains are λ in 90% of cases. The prognosis is better than that of typical myeloma, with a five-year survival of 60%.

Lymphoplasmacytic Lymphoma / Waldenstrom's Macroglobulinemia

LPL is a rare neoplasm composed of lymphocytes, plasmacytoid lymphocytes and PCs that involves the BM, lymph nodes, spleen and sometimes the PB. A characteristic IgM monoclonal protein is produced which can be detected in the serum (>3 g/dl), referred to as Waldenstrom's macroglobulinemia. Lymphomas with plasmacytoid morphology, as sometimes occur with B chronic lymphocytic leukemia, marginal zone lymphoma and follicular lymphoma are, by definition, excluded.

Clinical findings: Many of the clinical features are due to the M-protein. Hyperviscosity occurs in 10-30% and causes decreased visual acuity and increases the risk of stroke. Approximately 10% have neuropathies resulting from IgM activity to myelin sheath proteins. Some patients experience diarrhea secondary to IgM deposition in the gastrointestinal tract or bleeding caused by IgM binding with clotting factors or inhibition of platelet function. These IgM-related symptoms may be effectively treated with plasmaphoresis. LPL is sometimes associated with hepatitis C virus infection, and treatment of the hepatitis with interferon may cause regression of the lymphoma.

Laboratory findings: The serum M-protein expresses κ light chain in 75% of cases. Suppression of normal Ig production occurs in 50% of patients. Small amounts of light chain are often present in the urine.

PB and BM findings: The PB usually demonstrates rouleaux. Neoplastic cells are present in the PB in 30% of cases consisting of lymphocytes, plasmacytoid lymphocytes and occasional PCs with mild leukocytosis. Most patients have moderate to severe normochromic normocytic anemia and leukopenia, and thrombocytopenia are sometimes present. The BM demonstrates nodular and/or interstitial involvement with rare focal paratrabecular lesions. Dutcher bodies (nuclear pseudoinclusions) are a helpful feature for identifying plasmacytoid differentiation but are not entirely specific. The differential diagnosis of plasmacytoid BM infiltrates was discussed with PCM.

Flow cytometry: Surface Ig (sIg) is expressed, and some cells demonstrate cytoplasmic Ig (cIg), which is usually IgM, sometimes IgG, rarely IgA and never IgD. B-cell markers are present including CD19, CD20, CD22 and CD79a and CD38 is variably expressed. The neoplastic cells are negative for CD5, CD10 and CD23, which is helpful for the differentiation from B chronic lymphocytic leukemia (CD5+, CD23+), mantle cell lymphoma (CD5+) and follicular lymphoma (CD10+).

Genetics: Ig heavy chain genes are clonally rearranged. Rearrangement of the PAX-5 gene and the t(9;14) abnormality are present in 50% of patients.

Prognosis: LPL is an indolent disease associated with long survival.

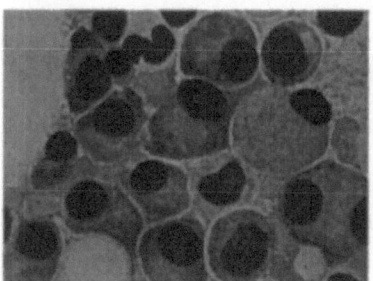

Figure 10.1. Plasma cell myeloma (BM aspirate, 500x).

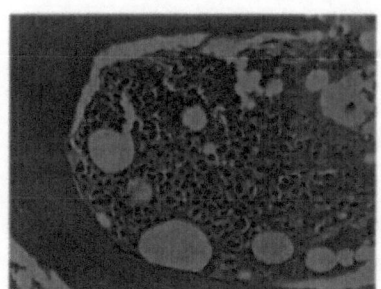

Figure 10.2. Plasma cell myeloma (BM biopsy, 100x).

10

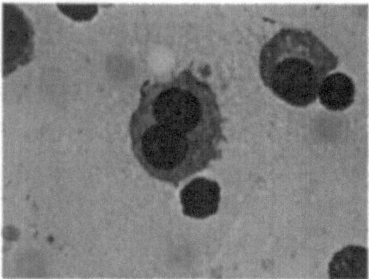

Figure 10.3. Plasma cells with Dutcher bodies (BM aspirate, 500x).

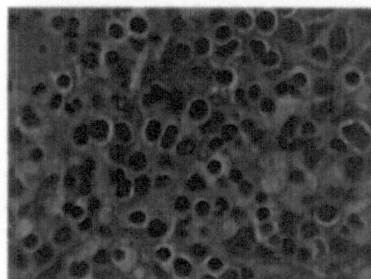

Figure 10.4. Inapparent plasma cell identifiable by Dutcher body (BM biopsy, 400x).

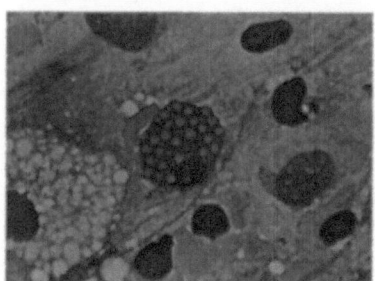

Figure 10.5. Mott cell (BM aspirate, 500x).

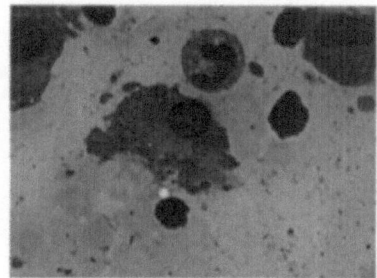

Figure 10.6. Flame cell (BM aspirate, 500x) sometimes seen in IgA myeloma.

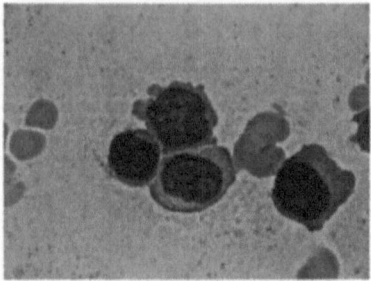

Figure 10.7. Blastoid morphologic variant of plasma cell myeloma (BM aspirate, 500x).

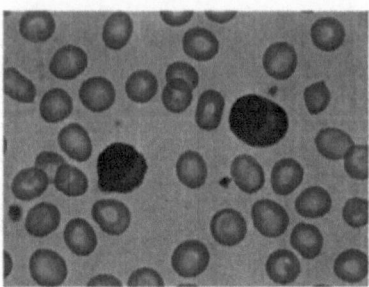

Figure 10.8. Plasma cell leukemia (PB, 500x).

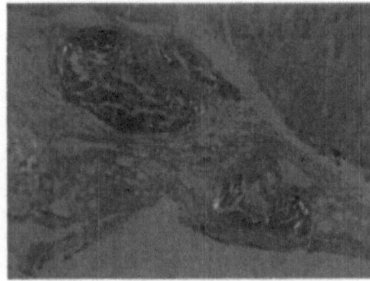

Figure 10.9. Amyloidosis of blood vessel (Congo red stain, polarized light, BM biopsy, 400x).

CHAPTER 11

Examination of the Bone Marrow for Lymphoma

Introduction

Bone marrow (BM) sampling for malignant lymphoma (ML) is done in a number of clinical settings:

- Staging of patients with diagnosis of ML already established from other tissue site.
- To attempt to establish an initial diagnosis of ML in a patient with lymphadenopathy in deep sites, not readily accessible for biopsy.
- Restaging of an ML patient to evaluate the effectiveness of treatment or for disease recurrence.
- Evaluation of a patient with cytopenia(s), fever, and/or organomegaly in whom ML is not specifically suspected.

The principles of BM examination for lymphoma are similar regardless of the indication, with some special considerations regarding Hodgkin lymphoma (HL), which is discussed separately at the end of this chapter. The following discussion is based on the perspective of BM examination for ML as a global process and not from that of individual diseases.

Peripheral Blood

Examination of the PB can yield valuable information in these cases. "Peripheralization," or "leukemic phase" of lymphoma, refers to the leakage of lymphoma cells into the PB. In the case of a dry tap, the PB may be the only source of neoplastic cells for immunophenoytping by FCM and may reveal useful morphologic features not apparent in the BM biopsy.

Bone Marrow Aspirate

Lymphoma is frequently not morphologically appreciated in the BM aspirate, because the neoplastic cells tend to blend with the surrounding non-neoplastic cells. Furthermore, focal fibrosis of the BM in the area of involvement by ML may mechanically hamper aspiration of these cells but not of adjacent normal hematopoietic cells. When seen, the appearance of ML cells in the aspirate will vary from small and mature-appearing to highly atypical malignant forms, depending on the type of lymphoma. The most important role of the BM aspirate is as a source of cells for immunophenotyping by FCM.

The BM biopsy is, by far, the most important specimen for the identification of ML. Bilateral BM biopsy sampling increases the diagnostic yield in staging cases. Some of the common types of ML cells are described in Table 11.1.

Bone Marrow: A Practical Manual, by Daniel A. Cherry and Tomislav M. Jelic.
©2011 Landes Bioscience.

Table 11.1. Appearance of common neoplastic B-lymphocytes in BM biopsy sections

Cell	Size	Nucleus	Cytoplasm	Diseases Seen In
Centroblast	Medium to large	Round or oval; vesicular chromatin; medium nucleoli, often pasted to nuclear membrane	Scanty; basophilic	FL and some cases of DLBCL
Centrocyte	Small to medium	Elongated and indented; dispersed chromatin; small inconspicuous nucleoli	Scant amount with indistinct cell borders	FL
Immunoblast	Large	Oval or round; areas of chromatin clearing; large, usually solitary, central nucleolus	Moderate amount with fairly distinct cell borders	DLBCL
Lymphoplasmacytoid	Small to medium	Round to oval; somewhat eccentric; reticular chromatin with or without clock face pattern	More basophilic and abundant than lymphocyte but less basophilic and abundant than PC; no prominent Golgi body	lymphoplasmacytic lymphoma/ Waldenstrom's macroglobulinemia; morphological variant of several types of lymphoma including SLL
Monocytoid B cell	Small to medium	Round or oval; occasional nuclear membrane irregularities; coarse reticular chromatin; small, usually solitary nucleoli	Abundant; sometimes clear	Marginal zone lymphoma
Plasma cell	Small to medium	Round; eccentric; reticular chromatin; may have inclusion	Abundant with prominent Golgi body	Myeloma, plasmacytoma; seen as reactive cell in several lymphomas, especially HL
Small lympho-cyte	Small	Round; occasional mild irregularities; coarse chromatin	Scanty; pale	SLL

DLBCL: diffuse large B-cell lymphoma; FL: follicular lymphoma; PC: plasma cell; SLL: small lymphocytic lymphoma.

11

A general rule of thumb is that low-grade lymphomas are higher-stage at the time of diagnosis and, therefore, are present in the BM more frequently than high-grade lymphomas at initial staging.

Patterns of Lymphoma Involvement in the BM Biopsy

There is a great deal of overlap of the architectural patterns of involvement of the BM biopsy by different kinds of lymphoma, but the pattern can still provide important clues. The most well known is the paratrabecular nodules of centrocytes with variable numbers of intermixed centroblasts seen with follicular lymphoma (FL). This pattern, however, like all others, is not specific. The vocabulary of lymphoma architecture in the BM is given in Table 11.2.

Paratrabecular refers to cells growing along, or hugging, the bony trabeculae. This pattern is always associated with fibrosis.

Intertrabecular involvement can be interstitial, nodular, or mixed interstitial and nodular. With interstitial involvement, ML cells are dispersed between fat cells or between normal hematopoietic cells. The latter is sometimes called monocellular, distinct from the pattern between adipocytes. Monocellular involvement may be minimal or dense with replacement of the hematopoietic cells. Sometimes small clusters of lymphoma cells form in association with interstitial involvement and may enlarge, called nodular reinforcement. Intertrabecular nodules may touch adjacent trabeculae, in which case the convex surface of the nodule is seen in contact with the trabeculum, as opposed the hugging pattern seen with paratrabecular involvement. The interstitial/monocellular, nodular and mixed patterns are sometimes further characterized as focal (patchy) or diffuse.

The **intrasinusoidal** pattern is rare and is specifically associated with anaplastic large cell lymphoma (ALCL), splenic marginal zone lymphoma (MZL) and hepatocellular T-cell lymphoma, although other patterns of involvement may be seen with these disorders. The intravascular variant of diffuse large B-cell lymphoma (DLBCL) is seen within blood vessels.

Lymphomas arranged in the interstitial and intrasinusoidal patterns are notorious for blending imperceptibly with normal surrounding cells and can be quite difficult to identify. Immunohistochemistry (IHC) is invaluable in these cases.

With **massive** involvement, the ML cells overrun the marrow architecture between trabeculae and may or may not extend to make contact with one or more trabeculae. Massive involvement may also be further characterized as focal or diffuse.

Table 11.2. Architectural patterns of BM biopsy involvement by ML

- Paratrabecular
- Intertrabecular
- Interstitial
- Nodular
- Mixed
- Intrasinusoidal
- Massive

Morphology of ML in the BM Biopsy

The cellular morphology of ML in the BM biopsy specimen may or may not be reflective of that seen in the lymph nodes (LNs) or other organs. Discordance occurs in approximately 25% of cases. This is commonly seen with FL in which the tumor grade, as determined by the relative number of centroblasts, appears to be lower than that seen in the LNs. Table 11.3 gives a summary of the patterns of BM involvement, cell type and incidence of BM involvement for several common lymphomas.

Immunophenotyping by FCM

The immunophenotype of the ML cells is fairly distinct for low-grade lymphomas and mantle cell lymphoma (MCL), and plays a critical role in making a specific diagnosis. This is less true for high-grade lymphomas in which the primary benefit of FCM is the determination of B cells versus T cells and demonstration of clonality.

FCM should be done in all cases of suspected ML regardless of the morphologic findings, except when there is a dry tap or in cases of HL. It is a good practice in all BM cases to collect extra specimen tubes to be held for FCM until the BM biopsy specimen has been examined. Even in cases in which ML is not suspected, lymphoid infiltrates are commonly seen unexpectedly. An ACD tube (yellow top) is preferred because of superior preservation of cell viability, but an EDTA tube (lavender top) can also be used.

Although FCM does not give an exact diagnosis, much can be inferred from the findings. Table 11.4 depicts the immunophenotypes of some of the more common B-cell lymphomas. One of the most frequently encountered difficulties is the differentiation of low-grade lymphomas, which may demonstrate much overlap in cellular morphology and pattern of involvement. This group includes the following:

- Small lymphocytic lymphoma (SLL)
- Follicular lymphoma (FL)
- Marginal zone lymphoma (MZL)
- Lymphoplasmacytic lymphoma (LPL)
- Mantle cell lymphoma (MCL)

Mantle cell lymphoma exhibits intermediate clinical behavior but is included because of the morphologic similarities to the others. Comparison of the expression of CD5, CD10 and CD23 (see Table 11.4) will render the answer in most of these cases.

It should be kept in mind that there are exceptions to all of the patterns given in Table 11.4. Sometimes IHC must be correlated with morphology and FCM. IHC for BCL-1, for example, will be positive in cases of MCL that are negative for CD5 or positive for CD23 by FCM.

With T-cell lymphomas, FCM can infer clonality by demonstrating an aberrant immunophenotype. Common examples include the following:

- Loss of CD7
- Co-expression of CD4 and CD8
- Absence of both CD4 and CD8

Of these, loss of CD7 is most common. There is some specificity to the patterns of antigen expression when correlated with the clinical finding, but less so than with low-grade B-cell lymphomas.

FCM is the most valuable technique for differentiating benign from neoplastic lymphocytic infiltrates. This is discussed more fully in the section of this chapter on the differential diagnosis of ML in the BM. FCM is indicative of clonal disease when there is

- Surface immunoglobulin (sIg) light chain restriction or the absence of SIg seen in B-cell lymphomas.
- Cytoplasmic immunoglobulin (cIg) light chain restriction in PC or lymphoplasmacytic neoplasms.
- Aberrant antigen expression in cases of T-cell lymphoma.

FCM can be falsely negative in some cases of ML, even when a seemingly adequate aspirate is tested. This is usually due to either focality of the disease or fibrosis in the area of involvement resulting in the aspiration of surrounding normal cells but not of the neoplastic cells. In some cases of large cell lymphoma, the neoplastic cells are too large to pass through the aperture of the FCM instrument.

Sometimes FCM is positive for ML that is not appreciated morphologically. This can be due to focality of the disease resulting in the presence of neoplastic cells in the portion of the specimen submitted for FCM but absent in the material submitted for microscopic examination. More commonly, this occurs due to the superior sensitivity of FCM as compared to microscopy in cases with a small tumor burden. Under ideal conditions, FCM can detect one neoplastic lymphocyte in 150,000 cells.

The significance of ML detected by FCM but not morphologically is not always clear, complicated by the fact that small numbers of "clonal" B cells can be seen in reactive conditions. Clinical correlation is crucial. In cases in which the diagnosis of ML has been established in another tissue, comparison of the immunophenotype is helpful. In many instances, the true significance cannot be determined and the phrase *monoclonal B cells of undetermined significance* (MBUS), may be used.

Keep in mind that FCM can be done on PB if the neoplastic cells are present. This can save the day when the BM is inaspirable or in cases when, for whatever reason, FCM was not done on the aspirate. It is much easier to have a patient drop back by the laboratory to give a blood sample than to repeat the BM sampling.

Immunohistochemistry

The usefulness of IHC is more variable than that of FCM. In practice, when the clinical picture, morphology and immunophenotype by FCM correlate neatly, IHC is superfluous. Common uses of IHC include the following:

- The differentiation of benign from neoplastic lymphoid infiltrates.
- The identification of ML involving the BM in an interstitial/monocellular and/or an intravascular/intrasinusoidal pattern.
- Demonstrating cIg light chain restriction in PCs or lymphoplasmacytic cells.
- Differentiating high-grade ML from metastatic carcinoma or melanoma.
- Differentiating ML from myeloid neoplasms.

Table 11.3. Characteristics of BM involvement by common lymphomas

Lymphoma	Pattern	Morphology	Fibrosis	Incidence	Notes
ALCL	Intravascular and/or interstitial; sometimes massive	Large anaplastic cells; wreath cells; Reed-Sternberg like cells	Localized to area of involvement	20%	Must do IHC for CD30 anytime ALCL is suspected
Burkitt Lymphoma	Patchy or massive	Monotonous; intermediate size cells; reticular chromatin with 2-4 inconspicuous nucleoli; basophilic cytoplasm; plasmacytoid and pleomorphic variants	Localized to area of involvement	50-60%	Necrosis common; starry sky pattern unusual in BM
SLL	Interstitial diffuse with or without nodular reinforcement; rarely shows proliferation centers	Small lymphocytes	Absent	75%	Almost never paratrabecular
DLBCL	Patchy or diffuse; makes contact with trabeculae	Centroblastic; immunoblastic	Variable	9-35%	See TCR and intravascular variants
FL	Almost always paratrabecular; rarely interstitial	Centrocytes with intermixed centroblasts	Localized to area of involvement	60%	Tumor grade seen in BM frequently lower than that seen in LNs
Hepatocellular T-cell lymphoma	Intrasinusoidal	Medium-size cells with oval nuclei; sometimes erythrophagocytosis	Absent	100%	May be subtle; IHC helpful
Intravascular DLBCL	Intravascular	Large atypical lymphocytes	Absent	See DLBCL	Possible association with erythrophagocytosis

continued on next page

Table 11.3. Continued

Lymphoma	Pattern	Morphology	Fibrosis	Incidence	Notes
LPL	Localized or diffuse interstitial with nodular reinforcement	Mixture of lymphoplasmacytic cells and PCs; <20% large cells (immunoblasts)	Frequent; localized or diffuse	80%	Occasionally paratrabecular; lacks proliferation centers
MCL	Intertrabecular nodules; 50% have paratrabecular aggregates; 70% have interstitial infiltrates	Mixture of small and medium size lymphocytes; mild to moderate nuclear irregularities; occasional large histiocytes without tingible bodies; blastoid variant with large atypical lymphocytes	Absent	75%	Bcl-1+ by IHC; occasionally paratrabecular; no proliferation or germinal centers
MZL, splenic, nodal and extranodal	Intrasinusoidal, early; interstitial and nodular, inter- or paratrabecular in more advanced disease; massive diffuse in late disease	Monocytoid B cells and PCs	Absent	10-40%	May have reactive germinal centers
PTCL, angioimmunoblastic	Patchy with localized interstitial; vascular hyperplasia	Medium and large atypical lymphocytes; reactive cells (PCs, eosinophils, epithelioid cells, B- immunoblasts)	Localized to area of involvement	65%	Prominent vascularity
PTCL, unspecified	Localized interstitial; sometimes paratrabecular	Small, medium or large lymphocytes with nuclear irregularities and pale cytoplasm	Localized to area of involvement	35-75%	
TCR-DLBCL	Patchy or massive	Few large B cells dispersed within background of many small to medium reactive T cells	Localized to area of involvement	See DLBCL	IHC is invaluable

ALCL: anaplastic large cell lymphoma; DLBCL: diffuse large B-cell lymphoma; FL: follicular lymphoma; LPL: lymphoplasmacytic lymphoma; MCL: mantle cell lymphoma; MZL: marginal zone lymphoma; PTCL: peripheral T-cell lymphoma; SLL: small lymphocytic lymphoma; TCR: T-cell rich.

11

Table 11.4. Immunophenotypic patterns of common B-cell lymphomas by FCM

	CLL/SLL	DLBCL	FL	LPL	MCL	MZL
CD5	+	-/+*	-	-	+	-
CD10	-	-/+**	+	-	-	-
CD23	+	-	+	-	-	-
CD43	+	+/-	-	-/+	+	-/+***
CIg	+/-	+/-	-	+****		-/+
SIg	+	+/-	+	+	++	+

*CD5⁺ in 10%; **CD10⁺ in 25-50%; ***splenic CD43⁻, nodal CD43⁻, MALT CD43⁺; ****IgM CIg⁺.

Demonstration of B- and T-cell compartmentalization by IHC: The ratio of T to B cells in benign aggregates is frequently roughly the same as that in the PB, or about 2:1, but may be perturbed in reactive conditions. The use of this ratio is inexact at best, but a marked predominance of one or the other, especially B cells, is evidence for ML. IHC markers that are helpful for showing lymphocyte lineage include the following:

- CD20 and CD79a for B cells
- CD3 for T cells

Note that UCHL-1 (CD45RO), a T-cell stain commonly used in other tissues, is absent from the list. UCHL-1 also stains granulocytes and is difficult to interpret in the BM biopsy. B-cell markers other than CD20 should be used when the patient has been treated with the anti-CD20 drug rituximab, for obvious reasons.

BCL-6 IHC may be helpful in differentiating benign aggregates from ML because it is usually negative in the former and positive in the latter.

The pattern of staining with **BCL-2** outside of the context of the follicular architecture of the LN is not helpful for identifying FL specifically, but diffuse positivity of lymphoid aggregates in the BM is indicative of B-cell ML in general.

IHC for identifying interstitial and intrasinusoidal involvement by ML: Involvement of the BM by ML in an interstitial/monocellular or intrasinusoidal/ intravascular pattern may not be evident without IHC, depending on the extent of involvement. Even when the ML is seen on the hematoxylin and eosin (H and E) stain, the extent of disease observed with IHC is often surprisingly more than previously appreciated. Neoplasms with a propensity for involvement in these patterns include the following:

- ALCL (intrasinusoidal and/or interstitial)
- Hairy cell leukemia (interstitial)
- MZL (intrasinusoidal)
- PTCL (interstitial)

IHC should be done in all cases of staging for these diseases or whenever one of these disorders is suspected, regardless of the BM appearance on the H and E stain. In cases of ALCL, CD30 should be used, because they can be negative for

Table 11.5. Hematologic cells that are negative for LCA

- ALCL cells (sometimes)
- PCs
- Reed-Sternberg cells
- Langerhans cells
- Follicular dendritic cells

leukocyte common antigen (LCA or CD45RA) and because the "null cell" cases are negative for CD3.

Kappa and lambda IHC for the demonstration of clonality: When PCs or lymphoplasmacytic cells are present, κ and λ IHC stains can be used to demonstrate clonality. The normal κ:λ is 2:1. Kappa light chain monotypism is usually defined as a κ:λ ratio of >10:1, and λ monotypism as <1:2. In practice, the predominance of one or the other in neoplasia is usually marked. Interpretation of cell surface staining for κ and λ is difficult and should not be attempted. Occasionally, one gets lucky and sees cytoplasmic light chain restriction in a ML that does not have plasmacytoid morphologic features, but this is uncommon and κ and λ IHC stains should not be done routinely in such cases.

IHC in cases of ML vs. metastatic carcinoma or melanoma: Lymphomas composed of large cells, most commonly DLBCL, can occasionally be difficult to distinguish from metastatic carcinoma or melanoma, especially if the pattern is intrasinusoidal. Carcinomas are positive for pan-cytokeratin such as AE1/AE3 and CAM5.2. Melanoma is positive for S-100, HMB-45 and MART-1.

IHC for differentiating ML from myeloid neoplasms: Large-cell ML can also be morphologically confused with myeloid malignancies, particularly in disorders with clustering of the myeloid cells, such as in some cases of acute myeloid leukemia (AML) and the phenomenon of abnormal localization of immature myeloid precursor cells (ALIP), which occurs in myelodysplastic syndromes (MDS). Myeloid and ML cells can be distinguished from one another by IHC for myeloperoxidase (MPO), which is positive in myeloblasts, or CD68, which is positive in monoblasts.

Differentiation of acute lymphoblastic leukemia (ALL) from ML by IHC: ALL will be positive for B- or T-cell markers, depending on lineage, but also expresses TdT by IHC. ML does not. The treatment of ALL is different from that of conventional B- and T-cell lymphomas, so this distinction is important.

Use of LCA (CD45RA) for identifying ML cells in the BM is generally discouraged. Because LCA is positive in a wide variety of white blood cells, interpretation is difficult. Furthermore, LCA is negative in some hematologic cells. All pathologists should know the short list of hematologic cells that can be negative for LCA (Table 11.5). A small IHC panel consisting of CD20, CD3 and CD30 will identify the vast majority of ML cells.

Molecular Studies (Including FISH)

Immunoglobulin heavy chain and T-cell receptor gene rearrangement studies by polymerase chain reaction (PCR) or Southern blot can be used to demonstrate clonality of lymphoid cells, and can be done on BM aspirate material collected in heparin (green top) tubes. PCR can be done on paraffin-embedded tissue blocks

Table 11.6. Specific chromosomal abnormalities in common lymphomas

Lymphoma	Chromosomal Abnormality
FL	BCL-2 gene: t(14;18)(q32;q21), present in 70-95%
MCL	BCL-1 or CYCLIN D1 gene: t(11;14)(q13;q32), present in 70-75%
Extranodal MZL	Trisomy 3 in 60%
	MLT gene: t(11;18)(q21;q21), present in 25-50%
Burkitt lymphoma	C-MYC gene: t(8;14)
	or
	Variants t(2;8) or t(8;22), seen in 100%

(if the tissue was not decalcified), but not if a mercurial based fixative such as B5 has been used.

There are chromosomal abnormalities that are useful for the diagnosis of specific lymphomas (Table 11.6). These can be detected by fluorescence in situ hybridization (FISH) using specific probes for the abnormality of interest. FISH can be done on BM aspirate material collected in heparin (green top) tubes or a BM clot section, but not on the decalcified BM biopsy.

Chromosomal Analysis (Cytogenetics)

Cytogenetics requires cells that are capable of achieving the metaphase stage of mitosis upon stimulation. Most ML cells are beyond the stage of maturation in which this is possible.

BM Changes Associated with Lymphoma

There are changes that can be seen in the BM with ML regardless of whether or not the BM is involved and regardless of changes that are related to involvement of the BM by ML.

Changes that may occur with or without involvement of the BM by the lymphoma include the following:

- Polyclonal plasmacytosis
- Reactive lymphoid nodules
- Granulomas
- Diffuse histiocytosis with phagocytosis
- Eosinophilic necrosis

The histiocytic phagocytosis is usually erythrophagocytosis and may be fatal. Eosinophilic necrosis is usually ischemic in nature and is seen with some cases of DLBCL.

Changes that occur specifically in association with BM involvement by ML include the following:

- Fibrosis
- Distention of sinuses
- Interstitial edema with or without hemorrhage (more common in post-treatment specimens)

Table 11.7. Morphologic features of benign lymphoid nodules in the BM

- Less than three seen in typical BM biopsy fragment
- Not paratrabecular
- Not associated with an interstitial infiltrate
- Not associated with fibrosis
- Occasionally may have reactive germinal centers*
- Have a mixture of B and T cells
- More heterogeneous with intermixed PCs and histiocytes
- Vascularity is usually prominent**

* More common in patients with autoimmune disease; may be seen in splenic MZL;
** May also be seen with PTCL.

Differential Diagnosis of ML in the BM

Reactive lymphoid nodules are commonly seen in the BM of elderly people and are of uncertain clinical significance. They may also be seen in patients with autoimmune disorders such as rheumatoid arthritis, systemic lupus erythematosis, autoimmune hemolytic anemia and idiopathic thrombocytopenic purpura. Reactive lymphoid aggregates, confusingly enough, can also be seen in the BM of a patient who has ML in another tissue without BM involvement. Several morphologic features are helpful for this differentiation (Table 11.7). The use of IHC and other studies for this purpose have been previously discussed.

Undifferentiated metastatic carcinoma and melanoma: Differentiation of these disorders from ML is usually accomplished by IHC (see previous discussion).

Hodgkin lymphoma (HL): HL involves the BM in a patchy or massive pattern and consists of Reed-Sternberg (RS) cells or mononuclear RS cell variants in a background of reactive lymphocytes, PCs and eosinophils. The RS cells can be identified by their IHC positivity for CD30 and, sometimes, CD15 and their negativity for pan-T-cell markers and LCA. About 20% of cases of HL demonstrate heterogeneous positivity of the RS cells for CD20, in contrast to the uniform, strong positivity seen in B-cell ML. Hodgkin lymphoma is not seen on FCM.

Acute lymphoblastic leukemia and acute myeloid leukemia may have an appearance similar to that of ML in a massive pattern or an interstitial pattern with local reinforcement. The differentiation from ML is easily accomplished with IHC, as previously discussed, as well as FCM.

Systemic mastocytosis (SM) may involve the BM in an interstitial or massive pattern. Paratrabecular involvement is common. Mast cells are spindle shaped with oval or reniform nuclei and abundant cytoplasm. The aggregates are frequently surrounded by a cuff of eosinophils. Mast cells are positive for tryptase and CD117 by IHC, and neoplastic mast cells are also positive for CD25.

Persistent polyclonal B-cell lymphocytosis is a rare disease seen almost exclusively in female smokers. This disorder is seen in the BM in an intravascular pattern composed of small binucleate lymphocytes. The lymphocytes are, as the name implies, polyclonal by FCM.

Differentiation of LPL from myeloma can, rarely, be problematic. The morphology may be similar, and both demonstrate cIg light chain restriction by

FCM and IHC. The monoclonal protein in LPL, as determined by serum or urine protein electrophoresis (SPEP, UPEP) and immunofixation electrophoresis (IFE), is IgM. By contrast, IgM is extremely rare in myeloma. PCM and plasmacytomas are typically BCL-1 positive while LPL is not. In most cases, the morphology, more lymphoid in LPL, and the presence of lytic bone lesions, seen with myeloma, make the proper diagnosis obvious.

Hodgkin Lymphoma (HL)

Staging of HL: BM involvement by HL is seen at the time of diagnosis in approximately 5 to 10% of patients. The practice of BM staging is controversial in patients who do not have cytopenias. The argument in favor of routine staging is that the presence of HL in the BM is of great prognostic significance, but this is questionable with modern therapy. The argument against routine staging is that it is an unnecessary discomfort and expense for the patient, since the BM is rarely positive. HL may be discovered in the BM as the initial site of presentation in HIV patients who do not have lymphadenopathy.

Morphology of HL in the BM: The pattern of involvement may be focal (30%) or diffuse (70%) and may be intertrabecular or paratrabecular. The infiltrate is heterogeneous, composed of sparse numbers of RS cells in a mixed inflammatory background composed of lymphocytes, PCs, histiocytes, neutrophils and eosinophils. The identification of RS cells may require the examination of multiple BM biopsy levels. RS cells must have the appropriate immunophenotype by IHC; that is, positive for CD30 (and sometimes CD15) and negative for B- and T-cell markers as well as LCA. Fibrosis is common. Necrosis may occur but is more common in post-treatment specimens. Hyperplasia of one or more hematopoietic cell lines and/or granulomas may be seen with or without BM involvement. HL cannot be accurately subtyped based on the BM appearance. HL is almost never appreciated in the BM aspirate, but milieu (increased PCs and eosinophils) may be seen.

Special rules for the diagnosis of HL in the BM: HL can be diagnosed in the BM if one of the following is present:
- RS cells with classic milieu.
- If RS cells have been identified in another tissue, then BM can have mononuclear RS variants with milieu.
- If HL has been proven in another tissue, then atypical cells with milieu is reported as "suspicious for HL."
- If HL has been proven in another tissue, then foci of fibrosis with typical or atypical RS cells is reported as "suspicious for HL."

Differential diagnosis of HL in the BM: The differential diagnosis includes ALCL, T-cell or histiocyte-rich DLBCL and PTCL. This is accomplished by IHC (see previous discussion). FCM may be helpful for identifying or ruling out other types of lymphoma, but it is not specifically useful for diagnosing HL. Flow cytometry should not be routinely done with BM staging of HL.

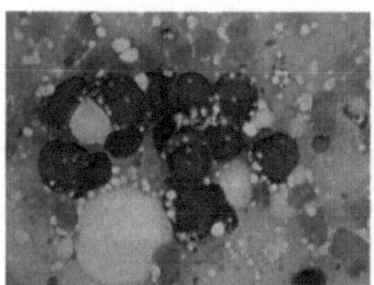

Figure 11.1. Burkitt lymphoma (BM aspirate, 500x).

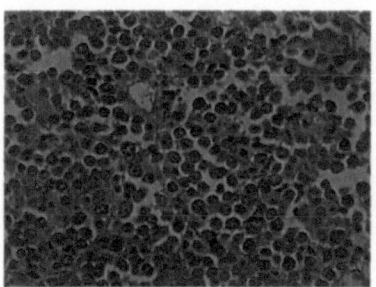

Figure 11.2. Burkitt lymphoma (BM biopsy, 200x).

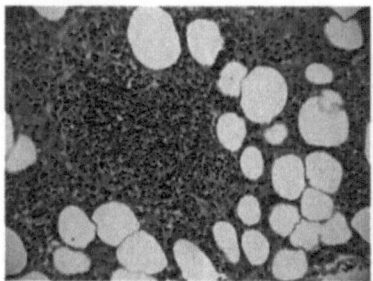

Figure 11.3. Small lymphocytic lymphoma (BM biopsy, 200x).

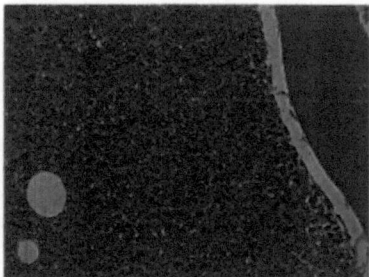

Figure 11.4. Diffuse large B-cell lymphoma (BM biopsy, 200x).

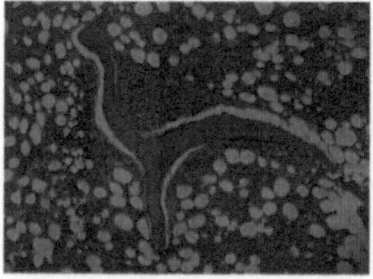

Figure 11.5. Classic paratrabecular pattern of BM involvement by follicular lymphoma (BM biopsy, 100x).

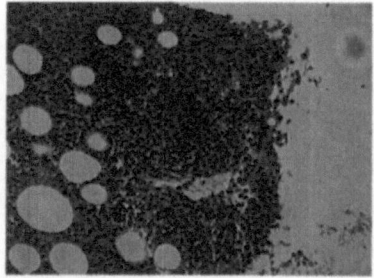

Figure 11.6. Lymphoplasmacytic lymphoma consisting of mature-appearing lymphocytes and plasmacytoid lymphocytes (BM biopsy, 200x).

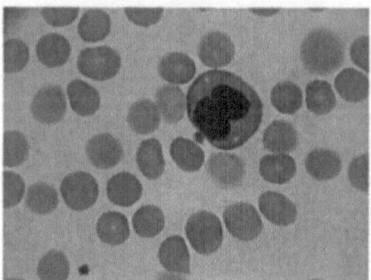

Figure 11.7. Mantle cell lymphhoma (cell with cleaved nucleus) in PB (500x).

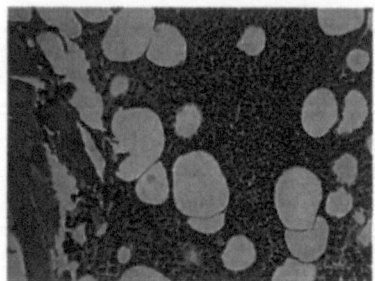

Figure 11.8. Mantle cell lymphoma (BM biopsy, 200x).

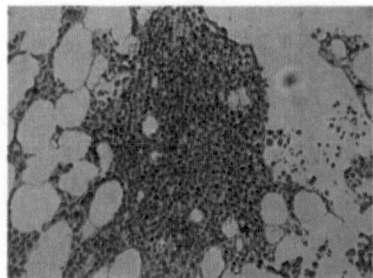

Figure 11.9. Cyclin D1 IHC showing nuclear positivity in mantle cell lymphoma (BM biopsy, 200x).

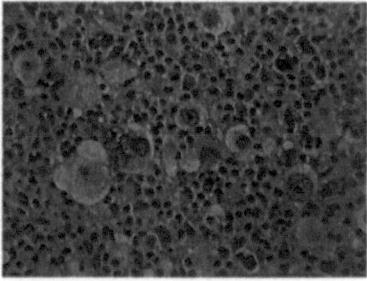

Figure 11.10. Hodgkin lymphoma, multiple Reed-Sternberg cells within a mixed inflammatory background (BM biopsy, 200x).

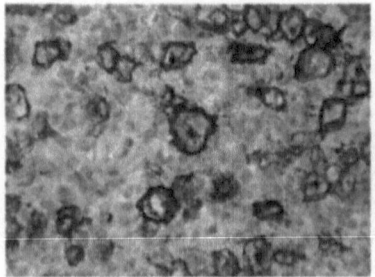

Figure 11.11. CD30 IHC showing golgi pattern positivity in Reed-Sternberg cells (BM biopsy, 200x).

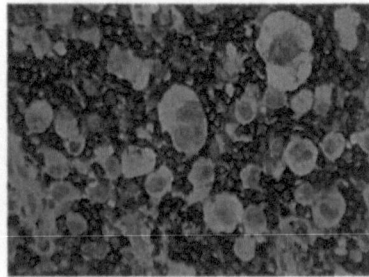

Figure 11.12. Absence of IHC staining for leukocyte common antigen in Reed-Sternberg Cells (BM biopsy, 200x).

CHAPTER 12

Evaluation of the Bone Marrow for Cytopenias

Introduction

The classification of quantitative abnormalities of red blood cells (RBCs), platelets and white blood cells (WBCs) is an everyday task encountered by all physicians, from family practitioners to neurosurgeons. Essential steps in evaluating these conditions encompass integration of clinical data and laboratory studies. The majority of cytopenias can be characterized by evaluation of results obtained by automated hematology instruments and by microscopic examination of the peripheral blood (PB) smear. In the minority of cases, such as when pathologic findings are present in more than one cell line (bicytopenia or pancytopenia), evidence of myelophthisis or suspicion for a hematologic neoplasm, BM examination may be necessary to establish a diagnosis. From the point of view of the community hospital pathologist, the majority of BM samples are submitted as part of the work-up of cytopenias with the explicit intention of identifying or ruling out neoplasia, hematologic or otherwise.

Neoplasms involving the BM are important causes of cytopenias but are not the most common. This chapter focuses on non-neoplastic cytopenias as well as general concepts for their evaluation and morphologic features. The discussion will center on BM morphology, rather than strictly follow a traditional encyclopedic listing and description of diseases. The limitations of BM examination for the diagnosis of these disorders are many and will be self-evident. A few diseases, such as iron deficiency anemia, will be presented in greater detail outside of the narrow context of BM morphology because of their ubiquitousness and importance in the overall context of human disease.

Anemia

Anemia is defined as a decreased concentration of hemoglobin that is usually accompanied by a decrease in the number of RBCs. The normal concentration of hemoglobin (Hgb) and number of RBCs depend on age, sex and altitude. At sea level, the normal Hgb is 14 to 18 g/dL for men and 12 to 16 g/dL for women. The normal number of RBCs is 4.6 to 6.2 million erythrocytes/μL for males and 4.2 to 5.2 million erythrocytes/μL for females.

The most commonly used classification schemes for anemia are based on pathophysiology (Table 12.1) and erythrocyte size (Table 12.2). Knowledge of both is useful because they are intimately entwined. There is imprecision in both and overlap between them.

Bone Marrow: A Practical Manual, by Daniel A. Cherry and Tomislav M. Jelic.
©2011 Landes Bioscience.

Table 12.1. Classification of anemia by pathophysiology

1. Anemia of blood loss
2. Anemia of decreased RBC production
   a. Iron deficiency
   b. Vitamin B12 and folate deficiency
   c. Copper deficiency
   d. Vitamin C deficiency
   e. Aplastic anemia
   f. Myelophthisis
      i. Metastatic tumor
      ii. Fibrosis
      iii. Granulomas
   g. Chronic kidney disease
   h. Hypopituitarism
   i. Protein depletion
   j. Myelodysplasia
   k. Parvovirus
3. Anemia due to increased RBC destruction
   a. Congenital hemolytic anemias
      i. RBC membrane defects
         1. Hereditary spherocytosis
         2. Hereditary elliptocytosis
         3. Hereditary stomatocytosis
         4. Hereditary pyropoikilocytosis
      ii. Abnormal hemoglobin (over 600 disorders described with some of the most common listed below)
         1. Beta thalassemia
         2. Alpha Thalassemia
         3. Hgb S
         4. Hgb C
         5. Hgb SC
         6. Hgb E
         7. Hgb E/β-thal
         8. Hgb SE
         9. Hgb S/β-thal
         10. Hgb Lepore
         11. Hgb M
      iii. RBC enzyme defects
         1. Glucose-6-phosphate dehydrogenase deficiency
         2. Pyruvate kinase deficiency
         3. Glucose phosphate isomerase deficiency
         4. Triosephosphate deficiency
      iv. Paroxysmal nocturnal hemoglobinuria

continued on next page

12

Table 12.1. Continued

b. Acquired hemolytic anemias
   i. Immune hemolytic anemias
     1. Autoimmune hemolytic anemias
       a. Warm-reactive antibodies
         i. Idiopathic
         ii. Secondary to SLE and other autoimmune diseases and viral infections (EBV, CMV, rubella, hepatitis, HIV)
         iii. Drug-induced (PCN, alpha-methyldopa, streptomycin)
       b. Cold-reactive antibodies
         i. Idiopathic cold agglutinin disease
         ii. Secondary to myeloproliferative disease and Mycoplasma pneumoniae infection
         iii. Paroxysmal cold hemoglobinuria (Donath-Landsteiner antibody), which may be idiopathic or secondary to syphilis or measles
     2. Isoimmune hemolytic anemias
       a. Hemolytic transfusion reaction
       b. Hemolytic disease of the newborn
         i. Rh incompatibility
         ii. Fetal-maternal ABO incompatibility
       c. Allograft-associated alloimmune hemolytic anemia
   ii. Hemolytic anemias due to mechanical damage of RBCs
     1. Prosthetic heart valves
     2. Thrombotic microangiopathy (DIC, TTP, HUS)
     3. Malignant hypertension
     4. Eclampsia and pre-eclampsia
     5. March hemoglobinuria
   iii. Hemolytic anemias due to direct effects of infection, chemical or physical agents
     1. Malaria
     2. Oroya fever (*Bartonella bacilloformis*)
     3. Clostridial sepsis
     4. Typhoid fever
     5. Miliary TB
     6. Sepsis in children caused by streptococcus, staphylococcus and meningococcus
     7. Spirochetes
     8. Thermal burns
     9. Some snake venoms
     10. Inorganic copper
     11. Arsine (ASH$_3$)

12

EBV: Epstein Barr virus; CMV: cytomegalovirus; HIV: human immunodeficiency virus; PCN: penicillin DIC: disseminated intravascular coagulation; TTP: thrombotic thrombocytopenic purpura; HUS: hemolytic uremic syndrome TB: tuberculosis.

Table 12.2. Classification of anemia by RBC size

1. Microcytic anemias (MCV <70-80 fL)
   a. Iron deficiency anemia
   b. Thalassemias
      i. Alpha thalassemia
      ii. Beta thalassemia
   c. Hemoglobinopathies (over 600 forms characterized)
   d. Sideroblastic anemia
   e. Chronic renal failure
   f. Lead intoxication
   g. Copper intoxication
2. Normocytic anemias (MCV 80-100 fL)
   a. Anemia of chronic disease (anemia of infection)
   b. Acute blood loss
   c. Hemolytic anemia
   d. Aplastic anemia
   e. Myelopthisic anemia
   f. Hematologic malignancy
   g. Endocrine diseases
   h. Chronic renal failure
3. Macrocytic anemias (MCV >100 fL)
   a. Vitamin B12 deficiency (MCV usually >120 fL all others usually 100-110 fL)
   b. Folate deficiency
   c. Alcoholism
   d. Chronic liver disease
   e. Myelodysplasia
   f. Aplastic anemia (usually normocytic)
   g. Arsenic, lead or chlordane intoxication
   h. Hypothyroidism
   i. Anorexia nervosa
   j. Pyridoxine-responsive megaloblastic anemia
   k. Erythroleukemia
   l. Reticulocytosis due to blood loss or hemolysis

Bone Marrow Morphologic Patterns Related to Pathophysiology of Anemia

In **acute blood loss**, there is usually not enough time for compensatory BM changes to occur. Bone marrow examination plays no role in the work-up of acute blood loss. If the blood loss becomes chronic, the picture is that of iron deficiency (see discussion below).

The cardinal feature of **anemia due to decreased production of RBCs** is BM hypocellularity, usually entirely due to erythroid hypoplasia with normal numbers of granulocytes and megakaryocytes. All cell lines are decreased in acquired aplastic anemia (AA). Lymphocytosis may also be present, including large numbers of hematogones, especially in children.

Table 12.3. Reticulocyte correction factors

Hct (%)	Correction Factor
40-45	1.0
35-39	1.5
25-34	2.0
15-24	2.5
<15	3.0

A decreased reticulocyte response, reflected by a decreased **reticulocyte count** in the PB, relative to the anemia, is reflective of decreased BM production of RBCs. Reticulocytes are non-nucleated RBCs that retain ribonucleic acid (RNA) and continue to produce hemoglobin. The RNA can be stained with methylene blue or brilliant cresyl blue, producing a dark reticulum network or dark blue granules. One thousand cells are counted, and the number of reticulocytes is reported as a percentage. For normal adults, the reticulocyte count is 0.5 to 1.5% or an absolute reticulocyte count (ARC = % reticulocytes x RBC count) of 24 to 84 x 10^9/L. Newborns have 2.5 to 6.5% reticulocytes, and this count falls to adult levels at about two weeks of age. Because RBCs lose their RNA approximately one day after entering the PB, the reticulocyte count reflects the rate of RBC production by the BM. There is, however, a complicating factor. As demand for RBC production increases, they are released into the circulation prematurely and the RNA may persist for two and one-half days or more. This is referred to as the reticulocyte left-shift and can result in overestimation of the reticulocyte count. The reticulocyte production index (RPI) can be used to correct for this error and is calculated as follows using the correction factors given in Table 12.3:

$$RPI = [(percent\ reticulocytes\ x\ Hct)/45] \times [1/correction\ factor]$$

An RPI less than 2 indicates decreased BM production of RBCs, and an RPI greater than 3 indicates an appropriate BM response to the anemia.

With **anemias of increased RBC destruction**, the BM is hypercellular due to erythroid hyperplasia. The erythroid precursor cells may show nuclear-cytoplasm dyssynchrony similar to that seen in megaloblastic anemia (see discussion below), but other findings such as granulocyte hypersegmentation and gigantism are lacking. Vitamin B12, folate and iron may eventually become depleted, causing a confusing appearance. The BM iron stain usually shows increased sideroblastic iron, sometimes with rings. Hemolytic anemia without stainable iron is suggestive of paroxysmal nocturnal hemoglobinuria (PNH). Bone marrow exhaustion and aplasia may occur eventually.

Characteristics of Anemias Based on RBC Size

The most common anemias in the United States in order of frequency are iron deficiency anemia, anemia of chronic disease (ACD) and megaloblastic anemia. These are the archetypes of the classification system based on RBC size.

Mean corpuscular volume (MCV) is the most useful measurement of RBC size. It is the most accurate of the RBC indices because it is obtained by direct measurement of the RBCs by the automated blood analyzer, rather than calculated. The normal range for the MCV is approximately 80 to 100 fL.

The classification of anemias based on RBC size is not absolute, since RBC size can vary for a particular etiology based on severity and other factors. For example, ACD is classically normocytic but, if severe, can be microcytic. AA is usually normocytic but may be macrocytic. Furthermore, when deficiencies of iron and vitamin B12 coexist, the MCV is normal.

Microcytic Anemias

Bone marrow examination is not indicated for the work-up of most microcytic anemias. Microcytic anemias are the most common, with iron deficiency and hemoglobinopathies accounting for about 35% of all anemias. Therefore, their morphologic manifestations are commonly observed in BM examined for other reasons. In general, the BM shows erythroid hyperplasia but not as pronounced as that seen with hemolytic anemias. The iron stain shows decreased storage iron in iron deficiency anemia, increased reticuloendothelial (macrophage) iron and decreased sideroblastic iron in ACD, often increased iron with thalassemias and ring sideroblasts in sideroblastic anemia. Bone marrow examination does play an important role in the diagnosis of sideroblastic anemia, the idiopathic form of which is a myelodysplastic syndrome (MDS), refractory anemia with ring sideroblasts (see chapter on MDS).

Iron Deficiency Anemia

Iron deficiency anemia (sideropenic anemia) is the most common in the United States and globally, accounting for about 30% of all cases. In fact, it is not hyperbole to state that iron deficiency anemia is one of the most important human diseases on our planet. Iron deficiency often coexists with other disorders that affect the BM and can confuse the BM and PB appearance as well as the clinical picture. The BM examiner, therefore, should have a strong fundamental knowledge of this disease. The diagnosis of iron deficiency anemia is established by clinical history, clinical laboratory parameters and PB examination.

In the United States and other developed countries, iron deficiency is most often caused by bleeding. In poor countries, it is most often caused by inadequate intake. Inadequate absorption is a rare cause. The etiologies of iron deficiency are enumerated in Table 12.4.

The diagnosis of advanced iron deficiency anemia is quite easy based on laboratory values and PB examination (Table 12.5), but these occur in sequence and all are not seen in early disease. As a reflection of decreased iron stores, decreased ferritin is the first laboratory abnormality to appear followed closely by increased transferrin (total iron binding capacity or TIBC), reflecting a compensatory increase in iron absorption. In the next phase there is a decrease in serum iron and transferrin saturation with an increase in serum transferrin receptor concentration. As storage iron becomes depleted, anemia develops, with normal RBC indices at first. The first RBC index to become abnormal is the RDW, as hypohemoglobinized cells appear in the PB followed by microcytosis and, lastly, decreased polychromasia.

12

Table 12.4. Etiologies of iron deficiency

1. Acute blood loss
   a. Hematemesis and melena from GI ulcers
   b. Massive nose bleed
   c. Trauma to large vessels
2. Chronic blood loss
   a. Uterine bleeding (dysfunctional, tumors)
   b. Benign and malignant GI tumors
   c. Inflammatory bowel disease (ulcerative colitis, Crohn's disease)
   d. Drugs that cause GI ulceration (ASA, other NSAIDs, glucocorticoids, KCL, ascorbic acid, ethacrynic acid)
   e. Drugs that alter hemostasis (warfarin, heparin, dipyramidole, clopidogril and others)
3. Inadequate iron intake
   a. Lack of meat in diet of poor people
   b. Strict vegetarians in developed countries
4. Inadequate iron absorption
   a. Chronic gastritis (Helicobacter)
   b. Partial gastrectomy or duodenectomy
   c. Celiac disease

Ferritin is the soluble form of storage iron (hemosiderin is the insoluble form) and serves as a readily available source of iron for any body requirement. Ferritin is present in the BM macrophages, nucleated RBCs (NRBCs), mature RBCs, hepatocytes, splenic macrophages and serum. Serum ferritin reflects BM iron stores and has a normal range of 30 to 300 ng/mL with a mean of 80 ng/mL in men and 49 ng/mL for women. Absolute Fe deficiency is indicated by a serum ferritin <20 ng/dL. The serum ferritin is increased when BM iron stores are increased in such conditions as ACD, sideroblastic anemia (including MDS), hemolytic anemias and hemochromatosis. Because ferritin is an acute phase reactant, it can be elevated in several disorders (Table 12.6). In the presence of inflammation or liver disease a serum ferritin <100 ng/dL is consistent with Fe deficiency. The best test for the diagnosis of iron deficiency is the soluble transferrin receptor assay.

Table 12.5. Laboratory findings in iron deficiency anemia

- Decreased Hgb
- Decreased MCV
- Decreased reticulocytes
- Decreased serum iron
- Decreased ferritin
- Increased transferrin
- Increased TIBC
- Decreased iron saturation
- Increased platelets

Table 12.6. Disorders associated with increased ferritin

- Infection and its recovery phase
- Post-operative state
- Ethanol abuse
- Neoplasms (Hodgkin lymphoma, acute leukemia, GI tumors)
- Acute and chronic liver disease
- Patients on hemodialysis

Transferrin is a β-1-globulin that transfers iron from duodenal mucosal cells to receptors on erythroblasts and transports unused iron to macrophages for storage. There is a compensatory increase in transferrin synthesis with iron deficiency anemia that is reflected by increased iron binding capacity (TIBC). The normal range for TIBC is 260-470 µg/dL. As opposed to ferritin, transferrin is decreased in acute phase conditions such as those listed in Table 12.6.

Soluble serum transferrin receptor levels normally range from 3.0 to 8.5 µg/mL and are increased in early iron deficiency anemia and normal or slightly decreased in ACD. The serum soluble transferrin receptor level is increased in iron deficiency before the Hgb drops.

PB findings: In the PB, when iron stores have been absent for some time, hypohemoglobinized cells (elliptocytes, microcytes) begin to appear, resulting in an increased RDW (poikilocytosis). An increased RDW is nonspecific as this is seen in many different anemias, but iron deficiency is unlikely if the RDW is normal. Due to the absence of iron, cytoplasmic maturation is delayed and the release of young RBCs into the PB is delayed, seen as an absence of polychromatophils (i.e., decreased reticulocytes). The platelets are slightly or moderately increased in the PB as a reactive phenomenon.

BM findings: In the BM, erythropoiesis is increased but ineffective. The late NRBCs are small, polychromatophilic (hypohemoglobinized) and have ragged cytoplasmic borders. The absence of iron stores on Prussian blue staining of the aspirate smear, core biopsy or clot section is the most sensitive and specific test for iron deficiency. If iron stores are decreased, then the core biopsy specimen may not be adequate for iron staining, since the decalcification process can remove small quantities of iron. Infants and children lack BM storage iron, so BM examination in these age groups is not helpful for the diagnosis of iron deficiency. Bone marrow examination is generally not needed but is sometimes done when the laboratory and PB findings are confusing, usually because of multiple etiologies for the anemia or when an underlying cause for the iron deficiency cannot be found.

The **differential diagnosis of iron deficiency** includes ACD, heterozygous α- and β-thalassemias and Hgb E. Knowledge of the duration of the microcytosis can be helpful since it will have appeared relatively recently with iron deficiency and ACD but will have been present since childhood with thalassemias and Hgb E. Anemia of chronic disease is the second most common anemia and is usually normochromic and normocytic but can be microcytic when severe and, therefore, is one of the most difficult anemias to differentiate from iron deficiency. Laboratory studies can be helpful for making this differentiation (Table 12.7). It should be kept

12

Table 12.7. Laboratory parameters in iron deficiency anemia versus anemia of chronic disease

	Ferritin	Serum Fe	RDW	Tfn Receptor Level	% Tfn Saturation	BM Fe
Fe↓	D	D	I	I	<10%	absent
ACD	I	D	N	N	>10%	I

Fe: iron; Tfn: transferrin; ACD: anemia of chronic disease; D: decreased; I: increased; N: normal

in mind that ACD and iron deficiency anemia often coexist. Anemia of chronic disease is discussed in more detail below.

Early iron deficiency anemia should be differentiated from heterozygous α- and β-thalassemias. In heterozygous thalassemias, the Hgb is slightly decreased or normal and the RBC count is normal or increased. The PB shows target cells and effectively increased erythropoiesis in the form of increased polychromatophilic cells (reticulocytes) and RBCs with basophilic stippling, features not seen with iron deficiency anemia. The thalassemias have increased iron stores and, thus, increased ferritin.

Hemoglobin E is common in Southeast Asians and is a more common cause of microcytosis in this population living in the United States. Homozygous Hgb E is associated with marked microcytosis (MCV ~60 fL) and a normal or slightly decreased Hgb. Heterozygotes have a normal Hgb and variable MCV that is usually slightly decreased but can be markedly decreased. The PB in Hgb E shows target cells without basophilic stippling or polychromatophilia. Bone marrow examination is rarely performed in thalassemias or Hgb E. In rare cases of Hgb E, hemolysis can occur with associated BM changes as previously discussed for anemias due to increased RBC destruction.

Anemia of Chronic Disease (Anemia of Inflammation)

Anemia of chronic disease is the second most common anemia worldwide, affects 2.5 percent of United States inhabitants and is the most frequent anemia in hospitalized patients. Anemia of chronic disease frequently leads to BM sampling and is a regular confounding presence in patients who undergo BM examination for unrelated reasons. The cardinal features are decreased serum iron and increased BM storage iron.

Multiple pathologic mechanisms contribute to ACD, but the main culprits are cytokines released from chronic inflammatory cells. Increased interleukin-6 upgrades the production of hepcedin, an iron-regulatory protein. Hepcedin causes the sequestration of iron in macrophages and decreased intestinal iron absorption. Cytokines released by chronic inflammatory cells also suppress erythropoietin (EPO) production, diminish the response of erythroid precursors to EPO and reduce mature RBC survival. In short, ACD is due to decreased availability of iron, decreased production of Hgb, decreased erythropoiesis and decreased RBC survival.

Clinical laboratory features of ACD include decreased serum iron and transferrin saturation. Percent transferrin saturation, however, is not decreased to the degree of that seen with iron deficiency anemia since serum transferrin is reduced in inflammatory diseases rather than elevated as with sideropenic anemia. Theoretically, serum transferrin decreases in inflammatory states as a defense mechanism since bacteria require Fe for growth.

PB findings: Examination of the PB demonstrates morphologically unremarkable RBCs. The anemia is normocytic and normochromic with a normal RDW but may become microcytic and hypochromic when advanced. Polychromatophilia and basophilic stippling are absent. The Hgb is rarely less than 9 g/dL.

BM findings: Bone marrow examination reveals normal or increased storage iron that is located predominantly in macrophages and a paucity of siderocytes.

ACD often coexists with iron deficiency anemia. Since serum ferritin is an acute phase reactant and is elevated in acute and chronic inflammatory states, the concentration of ferritin in these circumstances may not correctly reflect BM iron stores. However, a serum ferritin greater than 150 µg/L excludes iron deficiency, even in chronic inflammatory diseases. A good indication of iron deficiency is the presence of increased soluble transferrin receptors. In ACD, soluble transferrin receptors are normal or slightly decreased, reflecting suppressed erythropoiesis (i.e., reduced mass of RBC precursors). Laboratory tests useful for the differentiation of ACD from iron deficiency anemia are listed in Table 12.7.

In older persons, anemia is usually multifactoral and accompanied by iron deficiency. In these situations the laboratory picture can be confusing, and BM examination is advisable to assess iron stores and exclude extrinsic hematologic disease or myelophthisis.

Megaloblastic Anemia

Knowledge of megaloblastic anemia is extremely important to the BM examiner because the morphologic features can be similar to or indistinguishable from MDS or acute leukemia. Misdiagnosis can be tragic, leading to the patient receiving unnecessary, highly toxic treatment instead of the simple B12 and/or folate replacement needed to prevent irreversible neurological damage.

Both vitamin B12 and folate are necessary cofactors in a series of enzymatic reactions that result in the production of thymidine. Thymidine is one of the four constitutive bases of deoxyribonucleic acid (DNA); thus, deficiencies of B12 and folate result in decreased DNA synthesis and impaired nuclear maturation. Since thymidine is not present in RNA, where it is replaced by uridine, cytoplasmic maturation remains intact. The cardinal manifestation of B12 and folate deficiency is this dysynchrony of nuclear and cytoplasmic maturation.

Vitamin B12 is one of the essential vitamins and must be obtained from food. Red meat, seafood and some dairy products are good sources. Inadequate intake is seen only in strict vegetarians and in infants of strict vegetarians who are fed only breast milk. The neurological consequences for such infants are devastating.

In adults, B12 deficiency can also occur because of impaired absorption. Assimilation (absorption) of B12 is a complex multistep process, but a simplified review of the major elements suffices to understand the common mechanisms that block absorption. Pepsin and gastric acid help liberate B12 from food in the stomach. Liberated B12 binds with intrinsic factor, which, via receptors, enables absorption into the epithelial cells in the distal 60% of the ileum. Vitamin B12 in the blood is transported by transcobalamin 1 and transcobalamin 2. Enterohepatic circulation also occurs. The following obstacles may occur in the absorption process:

- Decreased intrinsic factor, pepsin and gastric acid due to autoimmune gastritis
- Inflammatory bowel disease (ulcerative colitis, Crohn's disease) and tropical sprue
- Ileal resection
- Bacterial overgrowth and fish tapeworm, which compete for B12
- Rare congenital deficiency of transcobalamin

Stores of vitamin B12 in the liver are large, and it takes years of curtailed absorption or intake for megaloblastic anemia to occur.

Folates are present in vegetables, cereals, fruits and dairy products. Inadequate intake is uncommon, and impaired absorption is rare, occurring in extensive gluten enteropathy and tropical sprue. Folate body stores are sufficient for about four months.

Laboratory findings: The most commonly used clinical laboratory tests for B12 and folate deficiency are serum B12 and folate levels; however, **homocysteine** and **methylmalonic acid** (MMA) are more sensitive, more specific and become abnormal earlier. Homocysteine is elevated in both vitamin B12 and folate deficiency, while MMA is elevated with B12 deficiency but normal with folate deficiency. Since these tests are superior, they should be done even when serum B12 and folate levels are normal. It is important to rule out B12 deficiency in all cases because folate therapy alone does not treat the neurological consequences of B12 deficiency.

PB findings: The earliest morphological finding of megaloblastic anemia in the PB is hypersegmentation, which appears even before the MCV increases. The criteria for hypersegmentation are five lobes in five percent or more of the neutrophils or any neutrophil with six lobes. This finding is highly specific and sensitive for megaloblastic anemia but not completely so, since hypersegmentation may occur as a manifestation of MDS or be so sparse in megaloblastic anemia that the examination of multiple smears is required to identify it. Additionally, the PB may contain band-form neutrophils and metamyelocytes with gigantism. The RBCs show prominent anisocytosis with oval macrocytes, teardrop cells, macrocytes and nonspecific poikilocytes as well as basophilic stippling and Howell-Jolly bodies. NRBCs are usually seen and demonstrate marked nuclear abnormalities, as described in the BM below. Polychromasia is absent (decreased reticulocytes). The MCV is usually greater than 110 fL and the RDW above 20. Thrombocytopenia and leukopenia may develop in advanced cases.

BM findings: Nuclear to cytoplasm maturation dyssynchrony is best seen in the erythrocyte precursor cells in the BM aspirate smears in which the nuclear chromatin is pale and fenestrated and the cytoplasm is fully hemoglobinized (pink). The BM is hypercellular with an inversion of the myeloid to erythroid ratio, an erythroid left-shift and cytomegaly. The erythroid nuclei, which are sometimes described as having a "pepperoni pizza" appearance, also show abnormalities of nuclear shape, binucleation, budding and multibudding. Karyorrhexis and mitoses are abundant. The myeloid cell line shows gigantism of the metamyelocytes and the bands as well as neutrophil hypersegmentation. A slight myeloid left-shift may occur, but myeloblasts are typically not increased. The megakaryocytes may be larger than normal and can be hypolobated or demonstrate complete separation of nuclear segments. Bone marrow iron is regularly increased. If the patient is transfused, then the erythroid hyperplasia lessens but the cellular abnormalities persist.

Differentiation of megaloblastic anemia from MDS: The morphologic features of megaloblastic anemia and MDS may be difficult or impossible to differentiate. In general, gigantism and hypersegmentation are more prominent in megaloblastic anemia, and a marked myeloid left-shift is indicative of MDS or AML. Flow cytometry (FCM) can be used to elucidate a myeloid left-shift and sometimes demonstrates aberrant markers or maturation suggestive of a myeloid neoplasm. The most conclusive evidence of MDS or AML is the presence of characteristic genetic abnormalities. If there is any doubt about the diagnosis, the clinical urgency of most cases will allow for a trial of vitamin B12 and folate therapy, which will result in an improvement of the erythroid cells in two to four days and the myeloid cells in 12 to 14 days in cases of megaloblastic anemia. Neutrophil hypersegmentation is the last morphologic change to disappear.

Thrombocytopenia

The normal platelet count is 100 to 400 x 10^3/μL, and the normal size is 2 microns with a mean platelet volume (MPV) of 7 to 10 fL. The platelet distribution width (PDW) is a measure of the dispersion of platelet sizes that is statistically calculated by the automated hematology analyzer and is analogous in meaning to the RDW.

Thrombocytopenia occurs for two reasons: the platelets are consumed and cleared from the blood and/or there is decreased production (Table 12.8). When thrombocytopenia is isolated and not accompanied by anemia or leukopenia, it is most likely that the platelets are being excessively consumed. In this setting, BM sampling is rarely necessary. If there is accompanying leukopenia or unexplained anemia, then BM examination is indicated to check for MDS, acute leukemia, BM fibrosis, or BM replacement by lymphoma, metastatic carcinoma, Gaucher's disease, etc.

When the number of platelets drops below 50 x 10^3/μL primary hemostasis becomes defective and may manifest as

- Easy bruising
- Gingival bleeding during tooth brushing
- Petechiae and purpura (bilateral) on the lower extremities and torso
- Bilateral epistaxis
- Menorrhagia
- Gastrointestinal and urinary bleeding

Thrombocytopenia is usually detected as an incidental finding on complete blood count (CBC) before it becomes symptomatic.

Table 12.8. Causes of thrombocytopenia by mechanism

1. Thrombocytopenia due to decreased production of platelets:
   a. Congenital amegakaryocytic thrombocytopenia
   b. Amegakaryocytic thrombocytopenia with radioulnar synostosis
   c. Thrombocytopenia with absent radii
   d. Acquired amegakaryocytic thrombocytopenic purpura
   e. MYH9 disorders
   f. Cyclic thrombocytopenia
   g. Non-immune drug induced thrombocytopenia
   h. Myelophthisis, including BM involvement with metastatic carcinoma, lymphoma, granulomas (TB, sarcoidosis, histoplasmosis) Gaucher's disease, leukemia
   i. MDS
   j. Myeloproliferative neoplasms (MPNs) excluding essential thrombocythemia
   k. Megaloblastic anemia
2. Thrombocytopenia due to increased consumption of platelets
   a. Immune thrombocytopenia
   b. Alloimmune thrombocytopenia
   c. Immune drug induced thrombocytopenia including heparin-induced thrombocytopenia (HIT)
   d. Disseminated intravascular coagulation (DIC)
   e. Thrombotic thrombocytopenic purpura (TTP)
   f. Hemolytic uremic syndrome (HUS)
   g. Eclampsia and pre-eclampsia
   h. HELLP (hemolysis, elevated liver enzymes and low platelets) syndrome
   i. Acute massive hemorrhage
   j. Acute infections

12

The first task for the pathologist is to determine whether the thrombocytopenia is real or false (pseudothrombocytopenia) due to laboratory artifact (Table 12.9). Pseudothrombocytopenia occurs in about 1 per 1,000 blood counts. A PB smear should be examined for platelet clumping or satellitism and to estimate the platelet count. Automated hematology analyzers may count small platelet clumps as one platelet or large platelet clumps as WBCs, or not count platelets attached to neutrophils or monocytes. Clumps of platelets are most commonly seen on the feathered edges or sides of the smear. The platelet count may be estimated using the previously described method (see Chapter 1, *Bone Marrow Basics*) or, alterna-

Table 12.9. Causes of pseudothrombocytopenia

- Platelet clumping due to agglutination with anti-CD41a (GPIIb/IIIa) auto-antibodies reactive in vitro at room temperature in the presence of EDTA
- Platelet satellitism
- Clot formation in blood collection tube
- Overfilling of vacuum blood collection tube (falsely elevated Hct and Hgb and falsely decreased WBCs and platelets)

Table 12.10. Diseases associated with pseudothrombocytopenia

- Hepatic cirrhosis
- Vasculitis
- Immunoglobulin A nephritis
- Antiphospholipid antibodies
- Cryoglobulinemia
- Viral infections
- Lymphoma
- Drug exposure

tively, by counting the platelets in a 1000x field and multiplying this number by 25. When the platelet count is normal, there are 9 to 12 platelets seen in a 1000x field. Pseudothrombocytopenia can also be diagnosed by analysis of scattergram patterns rendered by most modern automated hematology instruments, which will show platelet clumps. Unfortunately, false positives and false negatives are frequent by this method. Microscopic examination of a PB smear should still be performed. If the pseudothrombocytopenia is related to autoantibodies in the presence of ethylenediaminetetraacetic acid (EDTA, lavender top blood collection tube), then a specimen should be collected in a sodium citrate (yellow top) tube. If pseudothrombocytopenia cannot be ruled out on microscopic examination of the blood smear, then the gold standard is to collect blood by finger-stick into ammonium oxalate solution and count the platelets microscopically using a Burker or Neubauer chamber. Although pseudothrombocytopenia is a laboratory artifact, it can be associated with other diseases (Table 12.10).

PB findings: Clues to the cause of thrombocytopenia may be gleaned from examination of the PB smear, including platelet size (Table 12.11). When thrombocytopenia is due to increased consumption or clearance from the PB, the BM tries to compensate by releasing increased numbers of platelets into the PB, including young large or giant platelets. This is easily recognized in the PB as a mixture of various size platelets, normal, large and giant. The presence of young platelets in the PB can also be confirmed by counting reticulated platelets (young platelets containing RNA) by FCM. Platelets tend to be more monotonous and lack gigantism when the thrombocytopenia is caused by decreased

Table 12.11. Disease associations with platelet size

Increased platelet size (macrothrombocytopenia)
- MYH9 disorders (thrombocytopenias with inclusions in WBCs)
- Bernard-Soulier syndrome
- Gray platelet syndrome
- Mediterranean macrothrombocytopenia
- Montreal platelet syndrome
- X-linked thrombocytopenia with dyserythropoiesis/thalassemia

Decreased platelet size
- Wiscott-Aldrich syndrome

production. Neutrophils with toxic changes or atypical lymphocytes are clues of a recent infection, macro-ovalocytes and neutrophil hypersegmentation of megaloblastic anemia, and schistocytes of hemolytic microangiopathic disease (DIC, TTP, HUS).

BM findings: As previously stated, BM examination is typically not required to evaluate thrombocytopenia. An isolated decreased platelet count is usually treated as idiopathic thrombocytopenic purpura (ITP). If the platelet count does not respond to first-line treatment modalities for ITP (prednisone, intravenous immune globulin, anti-D), clinicians will sometimes perform BM sampling to rule out other etiologies before more toxic second-line drugs (cyclophosphamide, azathioprine, dexamethasone) are given. When splenomegaly is present, most clinicians will have the BM examined to exclude primary treatable BM disorders, principally MPNs, before splenectomy is performed. When increased consumption or clearing is the cause for the thrombocytopenia, the BM will show normal or slightly increased numbers of megakaryocytes and sometimes evidence of a megakaryocytic left-shift, such as increased size and hypolobation. In cases of decreased production, the BM will show megakaryocytic hypoplasia, generalized aplasia and fatty replacement, granulomas, neoplasia or evidence of the other diseases listed in Table 12.8.

Neutropenia

The normal number of neutrophils (absolute neutrophil count or ANC) in the PB is age and race dependent and for adult Caucasians and Asians ranges from 1.5 to 7.0 x 10^3/µL. Africans, African Americans and Arab Jordanians as well as African and Yemenite Jews have a lower ANC without propensity to infection and have a lower limit of normal ANC of 1.0 x 10^3/µL. In adult Caucasians and Asians neutropenia is arbitrarily designated as mild (1.0 to 1.5 x 10^3/µL), moderate (0.5 to 0.9 x 10^3/µL), severe (0.1 to 0.4 x 10^3/µL) and profound (less than 0.1 x 10^3/µL).

The causes of neutropenia in adults in order of greatest to least frequency are as follows:

- Drug reactions
- Infections
- Neoplasms replacing the BM
- Myeloablative therapies
- Pure white cell aplasia
- Aplastic anemia
- Hypersplenism
- Megaloblastic anemia
- Large granular lymphocytosis or LGL leukemia

When the ANC in adults drops below 0.5 x 10^3/µL, infections from commensal organisms present on the skin, gingiva, anal region and GI tract can occur, and when less than 0.1 x 10^3/µL life-threatening infections may develop. When profound neutropenia is prolonged the patient is at great risk of dying from Gram-negative sepsis or fungal infections. It is sometimes difficult to diagnose infection, since the neutropenia results in smaller inflammatory infiltrates and lessens the classic signs and symptoms of inflammation (tumor, dolor, rubor). For example, chest X-ray findings of pneumonia may be subtle and difficult to detect.

In younger children the lower limit of normal for the ANC is greater than that for adults. In neonates during the first three days of life, the normal ANC is 1.8 to 7.2 x 10³/µL. From about three days until several months of age, the lower limit of normal is 2.5 x 10³/µL. In newborn babies the causes of neutropenia in order of frequency are as follows:

- Infection
- Maternal hypertension or drug treatment
- Maternal antibody production
- Constitutional disorders (cyclic neutropenia, Kostman syndrome, Chediak-Higashi syndrome)

When the ANC is below 1.0 x 10³/µL in infants, stomatitis, gingivitis, and skin infections are prone to develop. When the ANC falls below 0.5 x 10³/µL, infants and young children are at risk for more serious infections, such as sepsis, pneumonia and cellulites.

In the evaluation of decreased neutrophils, the first step is to determine whether the neutropenia is real or false (pseudoneutropenia). Pseudoneutropenia may be caused by agglutination of neutrophils or rarely by overfilling of blood collection vacuum tubes. Agglutination of neutrophils is easily detected on examination of the PB smear. Aggregates range from 2 to 80 or more cells, and in some cases there is platelet satellitosis around the neutrophils. Pseudoneutropenia may be associated with antibodies (IgG, IgM) against neutrophil surface antigens, high-dose gamma-globulin therapy, antibodies acquired during herpes infection, pneumonia caused by *Mycoplasma pneumoniae*, hepatic disorders, vasculitis or colon carcinoma, but can also be seen in healthy individuals. In some cases the agglutination occurs at room temperature but not at 37°C, irrespective of the type of anticoagulant used in the collection tube. It may occur at either temperature if the blood was collected in an EDTA tube.

A fairly comprehensive list of conditions associated with neutropenia is presented in Table 12.12. Most are rare disorders and are usually diagnosed by findings other than neutropenia. In general practice, infection is the most important cause of neutropenia, which occurs because of toxins produced by the micro-organisms that suppress the BM and/or because of destruction of neutrophils at the site of infection.

BM findings: BM examination in cases of neutropenia is performed mainly to assess the integrity and degree of granulopoiesis and to exclude intrinsic BM diseases associated with neutropenia, primarily MDS, leukemia or marrow replacement with metastases or fibrosis. Some generalizations can be made regarding the association of BM findings with various disease states and are presented in Table 12.13. As with anemia and thrombocytopenia, the most important function of BM examination is to exclude neoplasia. In the majority of instances the neoplasm is self-evident; however, this is not the case with hairy cell leukemia (HCL) or LGL leukemia.

Table 12.12. Conditions associated with neutropenia

Infection due to:
- Bacteria (Gram negative sepsis, disseminated mycobacteria)
- Viruses (EBV, parvovirus B19, hepatitis viruses, HIV-1)
- Ricketsiae
- Protozoa

Immune-mediated disorders, including:
- Autoimmune neutropenia
- Alloimmune neutropenia
- Agranulocytosis
- Pure white cell aplasia
- Systemic lupus erythematosus
- Bechet syndrome
- Sjogren syndrome

Drugs:
- Dose-dependent BM toxicity (cytostatic anti-neoplastic drugs, phenothiazines, antibiotics, gold, colchicines)
- Idiosyncratic drug reactions in rare patients in whom immunologically mediated depression of granulopoiesis can occur (aminopyrine, antithyroid drugs, sulfonamides, quinidine, levamisole)
- Anti-folate affect on DNA synthesis (trimethoprim, methotrexate)

Neoplastic conditions:
- MDS
- Hairy cell leukemia (HCL)
- T-cell neoplasms (T-cell LGL leukemia)
- Metastases to BM

Hereditary disorder:
- Congenital cyclic neutropenia
- Severe congenital neutropenia
- Kostman syndrome
- Chediak-Higashi syndrome
- Myelokathexis
- Shwachman-Daimond syndrome
- Pelger-Huët anomaly

12

Table 12.13. General morphologic patterns in the BM related to causes of neutropenia

Morphologic Pattern	General Mechanisms	Associated Diseases	Comments
Markedly decreased or absent granulopoiesis	Defective proliferation	BM replacement by metastases or fibrosis	Cytologic atypia of the granulocytic cells is often seen
		Idiosyncratic drug reactions	
		Myeloablative drug or radiation therapy	
		Aplastic anemia	
		Constitutional neutropenias	
Granulocytic hyperplasia	Defective maturation	Megaloblastic anemia	Cytologic abnormalities of the granulocytes often marked
	Defects of survival	Myelokathexis	In myelokathexis apoptosis and pyknosis are prominent
		Infections	
		Immune mediated neutropenias	May be granulocytic left-shift with infections and immune causes
			Neutrophil phagocytosis is sometimes seen with immune mediated disorders

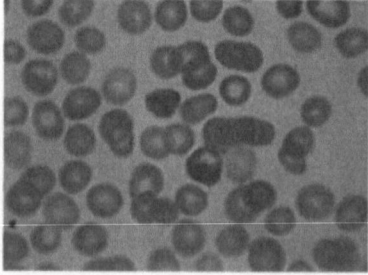

Figure 12.1. Polychromatophilic RBCs with basophilic stippling (PB, 500x).

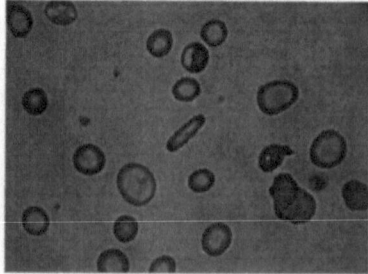

Figure 12.2. Iron deficiency anemia showing anisocytosis, microcytosis, hypochromasia and a pencil cell (PB, 500x).

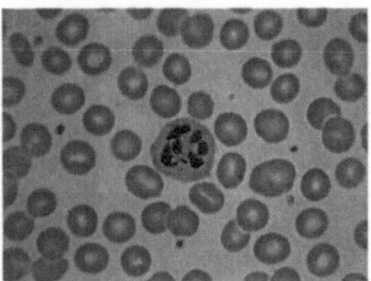

Figure 12.3. Megaloblastic anemia, neutrophil with gigantism and hypersegmentation (PB, 500x).

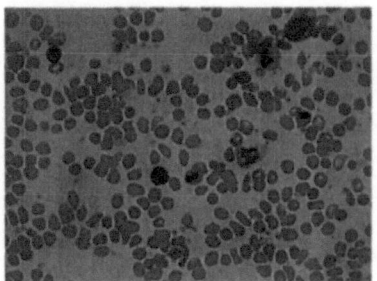

Figure 12.4. Megaloblastic anemia showing abnormal neutrophils, immature granulocytes, platelets and NRBCs (PB, 200x).

12

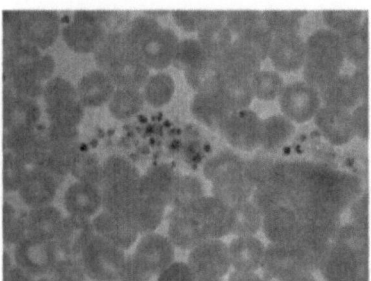

Figure 12.5. Platelet clumping resulting in pseudothrombocytopenia in the automated blood analyzer results (PB, 500x).

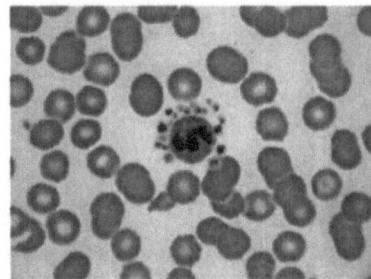

Figure 12.6. Platelet satellitosis (PB, 500x).

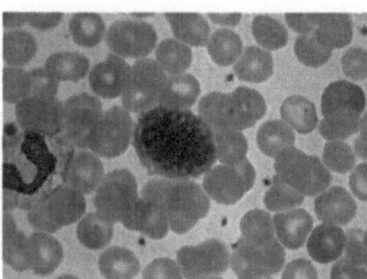

Figure 12.7. Giant platelet (PB, 500x).

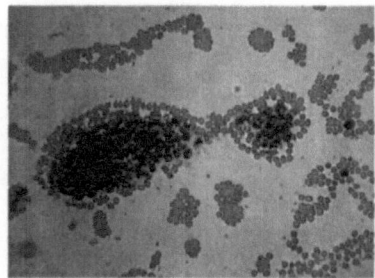

Figure 12.8. Neutrophil clumping resulting in pseudoneutropenia in the automated blood analyzer results (PB, 100x).

Ahmed A. Chronic idiopathic myelofibrosis, clinicopathologic features, pathogenesis, and prognosis. Arch Pathol Lab Med 2006 130:1133-1143.

Arber D, Thiele J, Tefferi A et al. 2007 Workshop of Society for Hematopathology and European Association for Hematopathology. Indianapolis: Indiana University School of Medicine, 2007. Conference Proceedings Handout.

Bagg A, Foucar K, Peterson L et al. Tutorial on Neoplastic Hematopathology. Boca Raton: Weill Medical College, 2008. Conference Proceedings Handout.

Bick B. Hematology Clinical and Laboratory Practice. Philadelphia: Mosby, 1993.

Binet JL. A new prognostic classification of chronic lymphocytic leukemia derived from a multivariate survival analysis. Cancer 1981 48:198-206.

Brunning R. Tumors of the bone marrow. In: Atlas of Tumor Pathology, 3rd ed. Washington, DC: AFIP Press, 1993.

Cao M. Polycythemia vera, new clinicopathologic perspectives. Arch Pathol Lab Med 2006; 130:1126-1132.

Carroll A. The t(1;22)(p13;q13) is nonrandom and restricted to infants with acute megakaryoblastic leukemia: a pediatric oncology group study. Blood 1991; 78(3):748-752.

Chang C. BCR/ABL-Negative chronic myeloproliferative disorders: JAK2 mutation and beyond. Arch Pathol Lab Med 2006; 130(8):1123-1125.

Craig F. Flow cytometric immunophenotyping for hematologic neoplasms. Blood 2008; 111(8):3941-3967.

Falini B. Cytoplasmic nucleophosmin in acute myelogenous leukemia with a normal karyotype. N Engl J Med 2005; 352:254-266.

Farhi D. Pathology of Bone Marrow and Blood Cells, 2nd ed. Philadelphia: Wolters Kluwer/ Lippincott Williams/Wilkins, 2008.

Foucar K. Bone Marrow Pathology, 2nd ed. Chicago: ASCP Press, 2001.

Frohling S. CEPBA mutations in younger adults with acute myeloid leukemia and normal cytogenetics: prognostic relevance and analysis of cooperating mutations. J Clin Oncol 2004; 22(4):624-633.

Garcia-Montero A. KIT mutation in mast cells and other bone marrow hematopoietic cell lineages in systemic mast cell disorders: a prospective study of the Spanish Network on Mastocytosis (REMA) in a series of 113 patients. Blood 2006; 108(7):2366-2372.

Greiner T. Diagnostic assays for the JAK2 v617f mutation in chronic myeloproliferative disorders. Am J Clin Path 2006; 125:651-653.

Hawley T. Flow Cytometry Protocols, 2nd ed. New York: Humana Press, 2004.

Howe R. The WHO classification of MDS does make a difference. Blood 2004; 103(9):3265-3270.

Hsi E. Hematopathology: Foundations in Diagnostic Pathology Series. Philadelphia: Churchill Livingstone Elsevier, 2006.

Jevremovic D. Molecular diagnosis of hematopoietic and lymphoid neoplasms. Hematol Oncol Clin North Am 2009; 23(4):903-933.

Karlovics R. A gain-of-function mutation of JAK2 in myeloproliferative disorders. N Engl J Med 2005; 352(17):1779-1790.

Khalil F. The criteria for bone marrow recovery post-myelosuppressive therapy for acute myelogenous leukemia, a quantitive study. Arch Pathol Lab Med 2007; 131:1281-1289.

Kjeldsberg C. Practical Diagnosis of Hematologic Disorders, 4th ed. Chicago: ASCP Press, 2006.

Knowles D. Neoplastic Hematopathology, 2nd ed. Philadelphia: Lippincott Williams and Wilkins, 2001.

Kolomietz, E. Primary chromosomal rearrangements of leukemia are frequently accompanied by extensive submicroscopic deletions and may lead to altered prognosis. Blood 2001; 97(11):3581-3588.

Lion T. The translocation t(1;22)(p13;q13) is a nonrandom marker specifically associated with acute megakaryocytic leukemia in young children. Blood 1992; 79(12):3325-3330.

MacPherson R. Henry's Clinical Diagnosis and Management by Laboratory Methods, 21st ed. Philadelphia: Saunders Elsevier, 2007.

Medeiros L, Fu K, Abruzzo L et al. Neoplastic Hematopathology Update, New Insights Into Old Questions. 2008. Fajardo, Puerto Rico: University of Nebraska Medical Center, 2008. Conference Proceedings Handout.

Nguyen D. Flow Cytometry in Hematopathology, A Visual Approach to Data Analysis and Interpretation, 2nd ed. New York: Humana Press, 2007.

Pardanani A. Systemic mast cell disease without associated hematologic disorder: a combined retrospective and prospective study. Mayo Clin Proc 2002; 77:1169-1175.

Pasqualucci L. Mutated nucleophosmin detects clonal multilineage involvement in acute myeloid leukemia: impact on WHO classificaton. Blood 2006; 108(13):4146-4155.

Peterson L. Acute myeloid leukemia with the 8q22;21q22 translocation: secondary mutational events and alternative t(8;21) transcripts. Blood 2007; 110(3): 799-805.

Rai KR. Clinical staging of chronic lymphocytic leukemia. Blood 1975 46:219-234.

Rice L. Current management of myeloproliferative disorders. Arch Pathol Lab Med 2006; 130:1151-1156.

Roumiantsev S. Distinct stem cell myeloproliferative/T lymphoma syndromes induced by ZNF198-FGFR1 and BCR-FGFR1 fusion genes from 8p11 translocations. Cancer Cell 2004; 5:287-298.

Sanchez S. Essential thrombocythemia, a review of diagnostic and pathologic features. Arch Pathol Lab Med 2006; 130:1144-1150.

Smith M. Mutation of CEBPA in familial acute myeloid leukemia. N Engl J Med 2004; 351(23):2403-2407.

Soupir C. Philadelphia chromosome-positive acute myeloid leukemia, a rare aggressive leukemia with clinicopathologic features distinct from chronic myeloid leukemia in myeloid blast crisis. Am J Clin Pathol 2007; 127:642-650.

Sun X. Comparative analysis of genes regulated in acute myelomonocytic leukemia with and without inv(16)(p13q22) using microarray techniques, real time PCR, immunohistochemistry, and flow cytometry immunophenotyping. Mod Pathol 2007; 20:811-820.

Swerdlow S. WHO Classification of Tumors of Haematopoietic and Lymphoid Tissues, 2nd ed. Geneva: WHO Press, 2008.

Tefferi A. Atypical myeloproliferative disorders: diagnosis and management. Mayo Clinic Proc 2006; 81(4):553-563.

Tefferi A. Bcr/abl-negative, classic myeloproliferative disorders: diagnosis and treatment. Mayo Clin Proc 2005; 80(9):1220-1232.

Tefferi A. Blood eosinophilia: a new paradigm in disease classification, diagnosis, and treatment. Mayo Clin Proc 2005; 80(1):75-83.

Tefferi A. Chronic myeloid leukemia: current application of cytogenetics and molecular testing for diagnosis and treatment. Mayo Clin Proc 2005 80(3):390-402.

Tefferi A. Classification and diagnosis of myeloproliferative neoplasms: the 2008 World Health Organization criteria and point-of-care diagnostic algorithms. Leukemia 2008 22(11):2118-2119.

Tefferi A. Proposals and rationale for revision of the World Health Organization diagnostic criteria for polycythemia vera, essential thrombocythemia, and primary myelofibrosis: recommendations fron an ad hoc international expert panel. Blood 2007; 110(4):1092-1097.

Tefferi A. The JAK2(V617F) tyrosine kinase mutation in myeloproliferative disorders: status report and immediate implications for disease classification and diagnosis. Mayo Clin Proc 2005; 80(7):947-948.

Vardiman J. The World Health Organization (WHO) classification of the myeloid neoplasms. Blood 2002; 100(7):2293-2302.

Wells D. Myeloid and monocytic dyspoiesis as determined by flow cytometric scoring in myelodysplastic syndrome correlates with the IPSS and with outcome after hématopoietic stem cell transplantation. Blood 2003; 102(1):394-403.

Wilkins B. Bone marrow pathology in essential thrombocythemia: interobserver reliability and utility for identifying disease subtypes. Blood 2008; 111(1):60-70

Wood B. 9-color and 10-color flow cytometry in the clinical laborarory. Arch Pathol Lab Med 2006; 130:680-690.

SR